Growth, Development and Reproduction

Dennis Taylor

Series editor
Fred Webber

CAMBRIDGE
UNIVERSITY PRESS

PUBLISHED BY THE PRESS SYNDICATE OF THE UNIVERSITY OF CAMBRIDGE
The Pitt Building, Trumpington Street, Cambridge CB2 1RP, United Kingdom

CAMBRIDGE UNIVERSITY PRESS
The Edinburgh Building, Cambridge CB2 2RU, United Kingdom
40 West 20th Street, New York, NY 10011–4211, USA
10 Stamford Road, Oakleigh, Melbourne 3166, Australia

First published 1996
Third printing 1997

Printed in the United Kingdom at the University Press, Cambridge

A catalogue record for this book is available from the British Library

ISBN 0 521 42203 5 paperback

Designed and produced by Gecko Ltd, Bicester, Oxon

This book is one of a series produced to support individual
modules within the Cambridge Modular Sciences scheme.
Teachers should note that written examinations will be set
on the content of each module as defined in the syllabus.
This book is the author's interpretation of the module.

Cover photo: Germinating coconuts (*Cocos nucifera*) on beach of
Pulau Tioman, West Malaysia (Harold Taylor ABIPP, Oxford
Scientific Films)

Contents

Acknowledgements

The author wishes to thank Dr Paul Davis for his valuable review and comments on the ethical issues of abortion.

Diagrams and tables

Figs 1.12 and 1.13 after Soper, R. and Smith, T. (1979) *Modern Human and Social Biology*, Thomas Nelson and Sons Ltd; table 2.2 from Leifert, C., Clark, E. & Rothery, C. (1993) 'Micropropagation', *Biological Sciences Review* 5, 31–5; fig. 2.6 based on George, E.F. and Sherrington, P.D. (1984) 'Plant propagation by tissue culture' in *Handbook of Commercial Laboratories*, Exegetics Ltd; fig. 3.5 based on Primrose, S.B. (1987) *Modern Biotechnology*, Blackwell Scientific Publications; fig. 3.13 after Green, N.P.O., Stout, G.W., Taylor, D.J. & Soper, R. (1990) *Biological Science*, Cambridge University Press; table 4.4 based on data from HMSO (Crown copyright is reproduced with the permission of the Controller of HMSO); fig. 4.26 from *The Guardian*; fig. 5.1 from Hendricks, S.B. and Borthwick, H.A. (1954) 'Photoperiodism in plants' *Proceedings of the 1st International Photobiology Congress*, 23–25; fig. 5.6 from Luckwill, L.C. (1952) 'Growth-inhibiting and growth promoting substances in relation to the dormancy of apple seeds', reproduced by permission of *Journal of Horticultural Science*, 27, 53–67

Photographs

1.1, Omikron/Science Photo Library; 1.2, 1.14, 2.2c, 2.2f, 2.9, 3.3, 3.5, 3.9, 3.11, 4.8, 4.10, Biophoto Associates; 1.9, Sinclair Stammers/ Science Photo Library; 2.2d, Mr J. Forsdyke/ Science Photo Library; 2.5, Dr C.E. Jeffree/Science Photo Library; 2.7, 2.8, 2.11, Nigel Cattlin/ Holt Studios; 2.10, J. Holmes/Panos Pictures; 4.17, Prof P. Motta/Dept of Anatomy/ University 'La Sapienza', Rome/Science Photo Library; 4.19, David Scharf/ Science Photo Library; 4.20, Petit Format/CSI/ Science Photo Library; 4.22, Petit Format/Nestle/ Science Photo Library; 4.25a, Peter Brooker/ Rex Features Ltd; 4.25b Patsy Lynch/Rex Features Ltd; 4.27, Lennart Nilsson, *A Child is Born*, Doubleday.

Growth and development

The fundamental activities of living organisms can be summarised as nutrition, growth, reproduction, respiration, excretion, sensitivity and, for some, locomotion. Two of these activities, namely growth and reproduction, will be the theme of this book. Growth is usually accompanied by development, so it is usual to study both together. **Development** is a change in form which is genetically programmed and may be modified by the environment.

What is growth?

Many people would probably think that it is easy to answer this question. It is worth trying your own definition before reading further. Probably the most straightforward answer is *an increase in size*. This is quite reasonable, but things are not always

this simple. Consider the following situations, for example:

■ a zygote (a cell formed by the fusion of two gametes) can divide to form a ball of cells with no increase in mass or volume *(figure 1.1)*. This seems to be *growth without an increase in size*. Or is it *development*, not growth?

■ a plant organ, such as a potato tuber, may increase in size, as measured by mass or volume, simply by taking up water by osmosis. This appears to be an *increase in size without growth*. The increase could easily be reversed by the tuber losing water again, as when a plant wilts.

So, the question remains, what *is* growth?

Biologists prefer a more detailed and rigorous definition of growth, such as **growth is a permanent increase in dry mass of living material**. By specifying *dry* mass and a *permanent* (irreversible) *increase* we can ignore short-term fluctuations in water content that are particularly common in plants. *Mass* is the best indicator of growth of the whole organism. But even the definition of growth as a permanent increase in dry mass can have its problems. Normally, the processes of growth and development go hand in hand, but it was noted

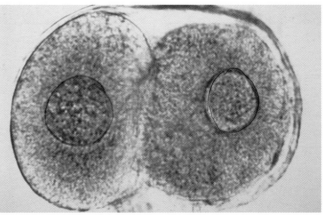

● **Figure 1.1** Light micrograph of a human embryo at the 2 cell stage. The zygote undergoes repeated mitotic divisions to form a ball of cells. The cells do not increase in size between divisions and therefore get smaller with successive divisions. Is this growth?

above that the early development of a zygote involves an increase in cell numbers by cleavage with no increase in mass or volume. Thus, according to our definition of growth, a zygote can develop without growing *(figure 1.1)*. A germinating seedling shows a net *loss* in *dry* mass until it starts to photosynthesise. By this time, much development, including that of a primary root and shoot, has taken place, accompanied by an increase in cell numbers, size and *fresh* mass. Should the definition specify that new dry mass must be living material? In some organisms, a significant amount of the dry mass added is of non-living material, such as xylem vessels in the wood of trees, and limestone in corals. Can the addition of non-living limestone be regarded as true growth?

Differentiation

In a multicellular organism, growth normally involves an increase in size as a result of cell division and a consequent increase in cell numbers. As the cells mature, they tend to specialise and become adapted for different functions, a process known as **differentiation**. Thus growth and development typically includes three separate processes, namely *cell division*, *cell enlargement* and *cell differentiation* (cell specialisation). The whole series of changes involved is controlled and regulated by interactions between the genes and the environment.

SAQ 1.1
What type of cell division is responsible for growth? mitosis

SAQ 1.2
The body cells of an adult are genetically identical. Suggest how cells with identical DNA can become specialised for different functions (differentiate).

Switching on and off of different genes at diff intervals @ time

Growth and development in plant meristems

Unlike the situation in animals, where growth occurs throughout the body, growth in plants is confined to particular regions called **meristems**.

The three processes of cell division, cell enlargement and cell differentiation are well illustrated by meristems because the processes are separated in time and space. This is particularly clear in root tips, which are responsible for the increase in length of the roots. Root tips are therefore a convenient model for studying growth in a multicellular organism. *Figure 1.2* is a longitudinal section (LS) through a root tip and shows three zones which, moving back from the tip, are the zones of cell division, enlargement (often elongation) and differentiation. This is a time sequence, the youngest cells being the dividing cells at the tip and the most mature being the differentiated cells. In animals, these three phases of growth also occur, but not in different locations, and so are not so obvious.

The zone of cell division in the root tip is a constant source of new cells while the root is growing. In the region of cell enlargement, cells take up water by osmosis and synthesise new materials, often becoming much longer *(figure 1.2)*. In the zone of cell differentiation, the cells become specialised for particular functions and develop specialised structures. For example, xylem vessels consist of long, dead tubes specialised for carrying water and mineral salts; phloem contains long, living tubes (sieve tubes) which carry organic solutes such as sugars; parenchyma is a type of packing tissue which in shoots may contain chloroplasts for photosynthesis (chlorenchyma); epidermal cells have a protective function and in roots may grow extensions, the root hairs, for water absorption. Remember that all these cells contain identical DNA and therefore identical sets of genes. Differentiation must therefore involve the switching on and off of different genes in different cells at different times. Cells destined to become xylem vessels, for example, must synthesise the strengthening material lignin, which reinforces their walls. The enzymes required for the relevant metabolic pathway are coded for by DNA. Each type of cell has its own unique spectrum of enzymes which controls its activities. The whole genetically programmed sequence of events which unfolds is known as **development**. The genetic

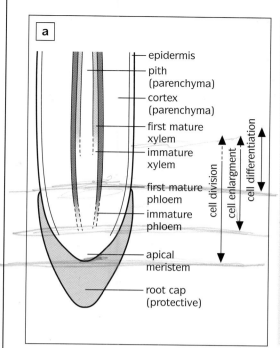

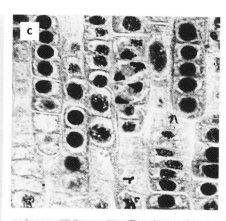

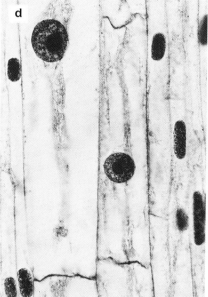

● *Figure 1.2*

a Simplified diagram of LS root tip.

b Photomicrograph LS root tip of *Vicia faba* (broad bean) (×100).

c Cells from zone of cell division (× 800).

d Cells from zones of cell elongation and differentiation (× 1500).

programme has evolved to respond to environmental stimuli, such as gravity, light and moisture, which help to control, regulate and direct growth. This means that development is a progressive series of changes which are a product of genes and environment. The root tip thus illustrates general principles which apply to all multicellular organisms.

Population growth

It is worth noting here the special case of unicellular organisms, such as bacteria, where cell division results in asexual reproduction and where we are more concerned with growth of the *population* than growth of individuals. Such population growth is considered later in this chapter.

Types of growth curve

Just as there is more than one way of defining growth, so there is more than one way of measuring it.

SAQ 1.3

From the preceding discussion, suggest at least **three** ways in which growth could be measured.

Absolute growth curves

A straightforward way of measuring growth is to plot size (for example length or height) or mass against time to produce a graph. Graphs which show increases in actual size with time are called **absolute growth curves** (or **actual growth curves**), because they show absolute (actual) growth. They are useful for showing *overall* growth patterns and *extent* of growth. Such growth curves often have a simple mathematical form, such as a straight line or, more commonly, an **S-shaped**, or **sigmoid**, shape

(1) increase in dry mass
(2) increase in complexity
(3) size

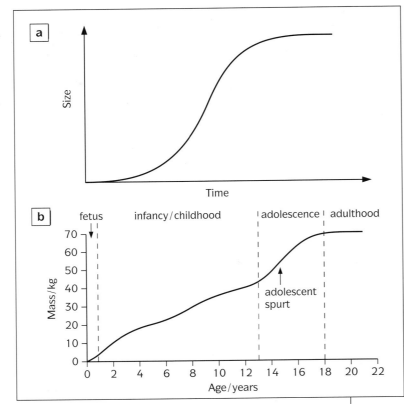

● Figure 1.3

a Idealised S-shaped absolute growth curve.

b Absolute growth curve for a sample of humans. Growth is measured as increase in mass. Four phases of growth are shown.

(figure 1.3a). Sigmoid curves are typical of many multicellular organisms, for example annual plants, insects, birds and mammals. Individual structures, for example fruits and leaves, may also show such curves. Later it will be seen that population growth in microorganisms also follows a sigmoid curve. Growth rate is slow at first but increases as the size of the organism (or population) increases. The larger the organism, the faster it grows, until a maximum rate is reached. This is maintained until cell division and expansion begin to slow down. Then the growth rate slows and eventually becomes zero.

Figure 1.4 shows the growth and development of a human female from age 5 to 19 years, and *figure 1.5* a human male also from age 5 to 19 years. Using the vertical scale provided on each figure, construct a table of height against age for the female and male. Then, using the data in the tables, draw graphs of height against age for the female and male. Draw the graphs between the same axes so that male and female can be compared. Compare your graphs with *figure 1.3b,*

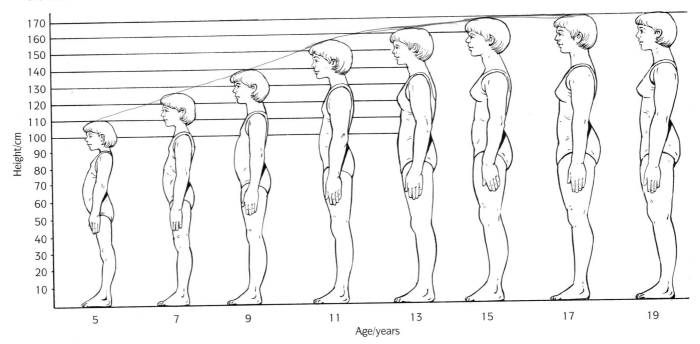

● Figure 1.4 The growth in height of an average human female from age 5 to 19 years. Note: there is a great deal of variation between individuals.

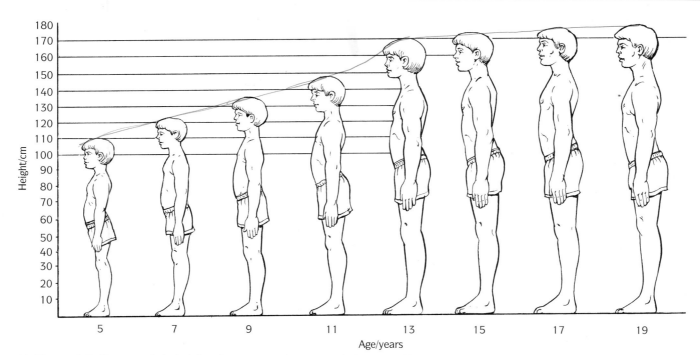

● **Figure 1.5** The growth in height of an average human male from age 5 to 19 years. Note: there is a great deal of variation between individuals.

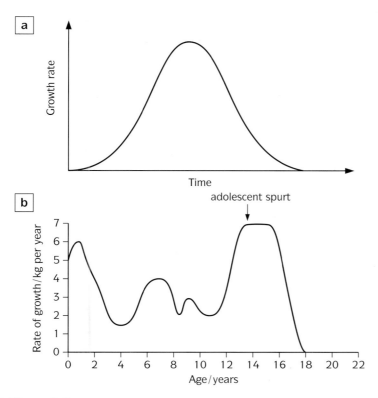

● **Figure 1.6**
a Absolute growth rate curve derived from the sigmoid curve shown in *figure 1.3a*.
b Absolute growth rate curve for a sample of humans measured as rate of increase of mass against time (age).

which is constructed in a similar way but shows the increase in *mass* for a sample of humans. In the case of height, the graph could also be formed by drawing a line from head to head in the equally spaced drawings of *figure 1.4*.

Absolute growth rate

Absolute growth rate is a measure of how the *rate* of growth varies with time. *Figure 1.6a* shows the typical bell shape obtained from the same data that produced the S-shaped growth curve of *figure 1.3a*. It shows growth *rate*. The peak represents the highest rate (corresponding to the steepest part of the S-shaped curve) and the final rate is zero (corresponding to the final flat part of the S-shaped curve). Using the data from the table you constructed above, calculate the increases in height per year for the human male and female and plot graphs of these against age.

Compare your graphs with *figure 1.6b*, which shows how the rate of increase in mass, rather than height, varies with time. Absolute growth rate graphs are useful for showing when growth is most rapid (the peak of the curve) and how the rate of growth changes with time (the slope of the curve). Be careful when interpreting such graphs to refer to growth *rate* and not to growth itself. A fall in the graph indicates a *slowing* of growth, not negative growth such as a loss in height or mass. A horizontal line indicates a *constant* rate of growth not zero growth. Thus in *figure 1.6b* the adolescent spurt eventually becomes a horizontal line once growth rate is constant. Note that at maturity the absolute growth rate is zero.

Relative growth rate

If a boy of 6 years old and one of 12 years old both grew 1 cm in one month, their *absolute* growth rates over this period would be the same. Yet common sense suggests that the 6-year-old boy is growing 'faster' because 1 cm represents a greater *relative* increase in height for him. **Relative growth rate**, also known as **specific growth rate**, can be measured by taking into account existing size. It can be calculated for height as follows:

$$\frac{\text{change in height (e.g. in one year)}}{\text{height at beginning of year}}$$

or

$$\frac{\text{absolute growth rate}}{\text{height}}$$

Other variables, such as mass, could be substituted for height. *Figure 1.7a* shows such a curve based on the S-shaped growth curve and *figure 1.7b* shows such a curve for a sample of humans. Note the high relative growth rate early in human life. Relative growth rate curves are useful for showing how *efficiently* an organism is growing, where **efficiency** means the growth rate *relative* to the size of the organism. Try to plot graphs of relative growth rate for the female and male using the data you have collected above.

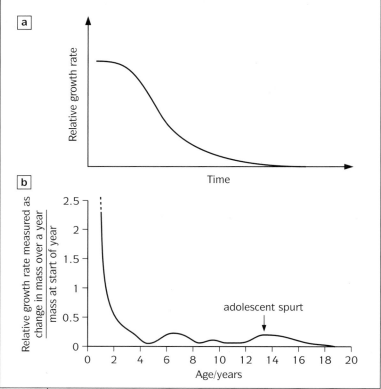

● **Figure 1.7**

a Relative growth rate curve derived from *figure 1.6a*.

b Relative growth rate curve for a sample of humans.

SAQ 1.4

Figure 1.8 shows changes in rate of growth and dry mass of a plant over a period of 40 days from the start of germination.

a Which graph, A or B, represents dry mass and which graph represents rate of growth?

b State one important way in which you would expect a graph showing increase in shoot length to differ from graphs A and B.

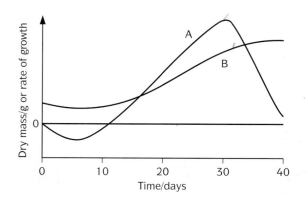

● **Figure 1.8**

c The rate of growth was calculated by measuring the change in dry mass over successive five-day periods and plotting the average change in mass per day. *A.* What types of growth curve are A and B? ~~Sigm~~ Actual (absolute)

d Describe the shape of graph B. Sigmad

e Why is there a negative growth rate immediately after germination? Plant uses stored food to grow.

Patterns of growth

The overall pattern of growth of the body of an organism may be **isometric** or **allometric**. With isometric growth, most parts of the body grow at more or less the same rate as each other throughout life, as in the case of fish and some insects. Thus a young fish has the same body *proportions* as an adult fish. With allometric growth, different parts of the body grow and develop at different rates. It is characteristic of mammals. In humans for example, the head of a baby is much larger relative to the rest of the body than in an adult. Thus allometric growth is accompanied by an obvious change in body shape.

Another variable in the *pattern* of growth of organisms is whether growth is **limited** or **unlimited**. With limited growth a maximum adult size is reached, after which size remains constant or may even decline. Annual plants, insects, birds and mammals show limited growth. Some organisms, such as woody perennial plants (trees and shrubs), fungi, algae, many invertebrates, fish and reptiles

● *Figure 1.9* Crop research. Wheat seedlings being grown in a growth chamber.

show unlimited growth. Here, growth never entirely ceases, although it does tend to slow with age.

Earlier we concluded that growth can be defined in various ways, such as an increase in size, cell number or complexity. We can now examine how these ideas can be applied in practice. Generally speaking, it is far easier to measure increase in size or mass than cell numbers or complexity.

Measuring growth in plants

The measurement and analysis of the growth of plants has important applications in agriculture. The better a crop plant grows, the higher its yield is likely to be. If techniques are available for measuring growth, then the effects of important variables, such as light, water and mineral availability, can be investigated, and optimum (ideal) growth conditions found. Experiments are commonly carried out in special growth cabinets in which environmental variables can be controlled and one variable at a time can be investigated (*figure 1.9*).

Annual flowering plants, such as cereals (for example wheat and barley) or the specially bred fast-growing *Brassica* species, make convenient material for studying growth in plants.

Increase in dry mass and other measurements

Measurement of mass is generally regarded as a better guide to growth than measurements of single dimensions such as height or length because it is more representative of the whole structure and is a better guide to the eventual yield. Fresh mass is relatively easy to determine, by weighing the whole plant. The problem of removing soil from roots can be avoided under laboratory conditions by growing the plants in nutrient solutions. Measuring dry mass is better than measuring fresh mass because the weight of the plant may be affected by fluctuations in water uptake and loss and such changes in weight do not represent true growth. However, dry mass measurements are more difficult and time-consuming to obtain (samples have to be oven-dried until they reach a

Time from sowing/days	Height/cm	Total dry mass/mg	Dry mass of endosperm/mg	Dry mass of embryo/mg
0		43	41	2
2		41	39	2
4	1.5	38	30	8
6	5.0	34	18	16
8	10.8	33	9	24
10	19.3	34	4	30
14	28.8	38		
21	43.2	50		
28	55.2	108		
35	62.6	201		
42	74.8	432		
49	89.3	865		
56	96.7	1707		
63		2765		
70		5803		

● *Table 1.1* Growth of oat plants

constant weight), and have the important disadvantage that they are destructive. A different sample of plants must therefore be used each time and the total number of plants required is much greater. Efforts must be made to make the samples truly representative, which ideally means taking the mean value of at least 30 individuals and ensuring that, as far as possible, all plants grow under identical conditions.

Table 1.1 provides data on the growth of oat plants from which you could draw a graph as in *figure 1.10* for up to 28 days, or a graph with different vertical scales for height and dry mass for up to 70 days. The endosperm is the food source

in cereal seeds. Before reading further you may like to try to account for the changes, over the first 10 days from sowing, in the dry mass of the endosperm and embryo and in the total dry mass.

The dry mass of the endosperm declines steadily over the first 10 days after sowing. As the seed germinates, the food in the endosperm (mainly starch) is digested by enzymes and the products of digestion move to the growing embryo. This accounts for the increase in dry mass of the embryo over the same period. The embryo is growing and requires nutrients such as glucose and amino acids.

Total dry mass declines slightly for the first eight days because aerobic respiration is taking place in the germinating seed. Carbon dioxide is a waste product of respiration and is lost as a gas, causing a loss in mass. The dry mass begins to increase once the gain in weight due to the photosynthetic activity of the first leaf starts to exceed the rate of weight loss as a result of respiration.

As the graphs show, one problem associated with using dry mass as a measure of growth is that growth may occur with only a little change, or even loss, in total dry mass. A further problem is that later in the growing season, the increase in dry mass slowly comes to a halt at a time when the seeds are developing and increasing greatly in size with stored food. This is caused by a diversion of nutrients from leaves to grain, the leaves dying in the process. By the criterion of change in dry mass, plant growth has stopped.

SAQ 1.5

Summarise the advantages and disadvantages of using dry mass as a measurement of growth.

Experiment to measure growth by changes in mass

In order to standardise conditions when carrying out experiments, seeds of the chosen plant, for example pea or wheat, can be

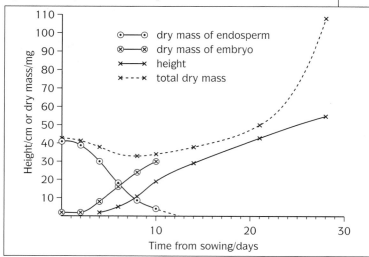

● *Figure 1.10* Growth of oat plants.

germinated in a commercial compost or in a sterile medium such as vermiculite, and watered with a standard culture solution containing the required mineral salts. Fresh mass can be measured after blotting dry the material to remove excess liquid. Various other measurements may also be recorded at the same time, such as leaf, internode and root length, plant height, and number or area of leaves. Dry mass can subsequently be determined by drying in a warm area for 24 hours, weighing, and repeating the procedure until the weight is constant. A suitable procedure might be to start with 300 seeds. Soak the seeds for 24 hours, take a sample of 30 for fresh and dry mass determination and plant the remainder. Further samples of 30 can then be taken at, for example, two-day intervals. When the seedling stage is reached the masses of whole plants, including roots, should be determined. Graphs of mean mass, or other measurements, against time can be plotted after 20 days.

Growth in length

Growth in length (the growth in one dimension of the whole plant or part of the plant) is likely to be easier to measure than changes in mass, and for many purposes it is just as useful. However, there are a number of drawbacks which generally make it a less satisfactory method. Both shoots and roots may vary in length according to environmental conditions. For example, shoots may grow longer when 'seeking' light without necessarily increasing in mass or cell number; similarly, roots may grow longer when 'seeking' water. Also, there may be more variation in linear dimensions than in mass. For example, two leaves could have the same mass but one may be shorter and wider. Mass is therefore more representative of true growth than length.

Experiment to measure growth by changes in length

A well-established method of measuring plant growth is to mark a growing root tip at equal intervals, about 1–2 mm apart, with Indian ink and then measure the distances between the marks at 24-hour intervals *(figure 1.11)*. Peas or runner beans have suitably large and straight primary roots (first

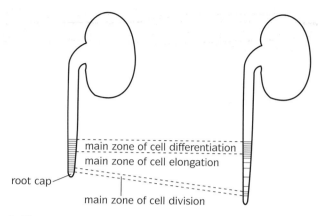

● **Figure 1.11** Demonstration of growth in a root.

roots). Starting from the tip, a primary root about 2 cm long can be marked with an ink-soaked cotton thread stretched taut between the ends of a bent piece of wire. The peas, runner beans or other large seeds can be pinned near the tops of pieces of softboard covered with filter paper, the base of each board placed in a beaker of water and the whole covered with polythene. Overall changes in length, and percentage increases in length for each interval on the primary root, can be plotted on graphs. The various types of growth curve already discussed on pages 3–6 can then be plotted.

Since dividing cells are found only in the root tip, elongating cells just behind the tip and differentiating cells still further from the tip, the marks in the elongation zone become further apart and those in the differentiation zone stay approximately the same distance apart *(figure 1.11)*. Note that a full definition of growth might include both cell division and cell differentiation as well as cell elongation, and this technique only measures cell elongation. More elaborate techniques involving microscopy are used to measure numbers of cells and degree of differentiation.

Measuring growth in insects

Measuring the growth of animals presents a new set of problems compared with plants. Much of our detailed knowledge has come from farm and laboratory animals because trapping wild animals and recording measurements from them is difficult. Insects, however, are useful animals for growth studies as they are convenient to keep and grow relatively quickly. They also show an interesting split into two groups with very different growth characteristics.

Metamorphosis

Most insects undergo metamorphosis during growth. **Metamorphosis** is a change in form from a larval stage to an adult stage during the life cycle. It is also characteristic of amphibians , as in the change from tadpole (larva) to frog (adult). During metamorphosis the body is re-modelled, which involves the breakdown of existing tissues by enzymes released from lysosomes, and the formation of new tissues. It is controlled by hormones.

Insects show two types of metamorphosis, known as complete and incomplete metamorphosis *(figure 1.12)*.

- With **complete metamorphosis**, a total change of form takes place with the result that the adult is very different in form from the larva.
- With **incomplete metamorphosis**, there is a gradual change in form from larva to adult. The larva undergoes a series of moults as it grows. Each successive stage is called a **nymph** or **instar** and is larger and more like the adult, though only the adult has functional wings.

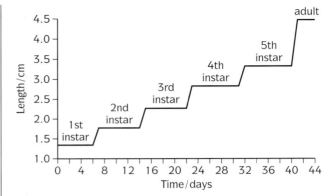

● **Figure 1.13** Growth curve based on body length for the short-horned grasshopper.

If you were trying to measure the growth of an insect showing incomplete metamorphosis, such as a locust or a grasshopper, an obvious method might be to measure body length at regular intervals. However, growth curves based on length have an unusual appearance *(figure 1.13)* because growth appears to take place in a series of spurts, with no growth in between. This is very misleading, because a growth curve based on the increase in dry mass of the insect is a smoother, typically S-shaped, or sigmoid curve *(figure 1.3a)*. Such growth starts more or less exponentially (see bacterial growth on page 13), then begins to slow down and finally ceases. Thus the *true* growth of the insect, based on increase in living material, is

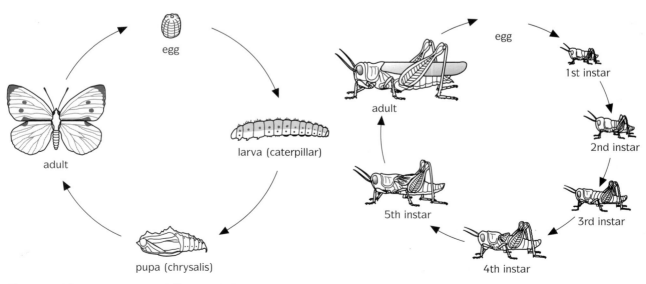

complete metamorphosis e.g. cabbage white butterfly

incomplete metamorphosis e.g. locust

egg

larva (caterpillar)

adult

pupa (chrysalis)

egg

1st instar

2nd instar

3rd instar

4th instar

5th instar

adult

● **Figure 1.12** The two types of life cycle in insects, complete and incomplete metamorphosis.

continuous and does not have the interrupted pattern shown in *figure 1.13*. The reason for the stepped growth curve for changes in body length is that the exoskeleton is too rigid to allow expansion. Growth in length is therefore confined to brief periods after moulting of the old exoskeleton, when a new exoskeleton is forming.

SAQ 1.6

Suggest **two** ways in which the daily growth of an insect could be measured, other than by determining increase in body length or dry mass.

SAQ 1.7

A representative sample of locust instars (of different stages) and adults was collected and the lengths of three features accurately measured. Mean lengths revealed the following features of growth:

head – maximum rate of growth early in development;

tibia (part of the hind leg) – steady rate of growth throughout development;

wings – maximum rate of growth late in development.

How might the changes observed be related to the life cycle of the insect?

Measuring growth in microorganisms and populations

Growth can be studied at any level of biological organisation, from cells, organs and organisms to populations and communities. A **population** is a group of organisms of the same species living together in a given place at a given time that can interbreed with each other. The term could refer, for example, to all the badgers living in an oak wood, or to all the bacteria living in a test-tube. Studying the growth of populations is important for a number of reasons. In agriculture, for instance, it is useful to know about the growth of pest populations. This might allow more effective timing of spraying with pesticides. The growth of human populations is another major concern. If we are to control human population growth, for example, then the effects of factors such as contraception, elimination of disease and economic development

on population growth must be investigated. A lot can be learnt by studying very simple organisms, such as yeasts or bacteria. These are unicellular, and have the advantage that large numbers can be grown relatively quickly in laboratories.

In a typical experiment, a small number of such microorganisms is introduced (inoculated) into a suitable nutrient medium and their increase in numbers is monitored over a period of time.

Both bacteria and yeast, a unicellular fungus, are commonly used. The nutrient media used will generally allow the growth of a wide range of microorganisms, so contamination must be prevented. It is therefore usual to use **aseptic techniques**, which involve using sterilised apparatus and materials. A description of these techniques is outside the scope of this book, but they are discussed in *Microbiology and Biotechnology* (page 50) in this series. Some of the principles involved in measuring growth and some of the problems commonly encountered will, however, be described here. Two types of count may be used: **viable counts** (living cells only) and **total counts** (living plus dead cells).

Viable counts

Viable counts are used when it is important to know the number of *living* microorganisms. For example, the effect of pasteurising milk could be investigated by finding the total number of living bacteria in samples of milk taken before and after pasteurisation. The basic principle on which the most common method of viable counting of bacteria is based is that, given a suitable medium in which to grow, each bacterium will multiply over one or two days to produce one visible colony. In practice it is usual to add a known small volume (for example $0.01\,cm^3$) of the sample to melted nutrient agar jelly in a petri dish or horizontal tube. This is rotated to mix, and the agar allowed to set. The culture is incubated as appropriate (for example at $30\,°C$ for 24 hours) and the number of colonies counted. In theory, the number of colonies should equal the number of bacteria in the original sample, but there may be so many colonies in the petri dish that it is impossible to count them.

Serial dilutions may therefore be necessary to get an appropriate number of colonies that can be counted easily. Replication (setting up two or more cultures) ensures more accurate results by allowing estimation of the amount of variation. Controls with no bacteria should also be set up to ensure that the technique is not allowing contamination by other bacteria.

Another method of viable counting is to measure a product of metabolism such as a gas (for example from respiration) or an acid. Such a method could be used for yeast as well as bacteria.

There are several problems associated with viable counting.

■ Aseptic procedures require special apparatus and techniques. Contamination occurs easily.
■ If more than one type of bacterium is present in a sample, as in milk, the culture conditions will not favour them all equally.
■ Bacteria are rarely distributed evenly throughout a sample. They are often found in clumps of variable numbers, so a single colony could be derived from many bacteria. It is therefore difficult to obtain reproducible results. However, high levels of accuracy are not often needed.
■ Some bacteria are pathogenic (cause disease). Care is therefore required with handling, and there are restrictions on the bacteria that can be grown in schools or colleges.

Total counts

Unlike viable counts, which measure only living microorganisms, total counts measure both living *and* dead cells in the sample. A number of methods may be used to obtain total counts. They can be used for bacteria and yeast. **Direct counting** using a microscope is possible, and for this a special slide known as a **haemocytometer slide** is useful. It has a ruled grid, usually $1\,mm^2$ in area and divided into 400 small squares. It is designed so that a known volume of sample covers the grid. A representative sample of cells can thus be counted and estimates can be made of the number in the total sample. In the case of yeast, the use of a stain which only stains dead cells, such as methylene blue, may

allow a viable count to be made. Direct counting has several disadvantages.

■ The technique requires practice.
■ The slide must be scrupulously clean.
■ It is difficult to distinguish bacteria from other small particulate matter.
■ Bacteria are often not evenly distributed throughout a sample, so small samples may result in large errors.

Another common method for obtaining a total count depends on the fact that the more bacteria there are in a solution, the more turbid (cloudy) it appears. **Turbidity methods** measure the amount of light that is transmitted through a suspension of the bacteria. If necessary, actual numbers can be obtained by comparing the results of unknown samples with those from samples containing known numbers.

Figure 1.14 shows the typical growth curve of a bacterial population. Under ideal circumstances, some bacteria can grow and divide as frequently as once every 20–30 minutes. At this rate, though, the mass of organisms would exceed that of the planet within a few days, clearly an impossible achievement. Instead, there are always factors limiting population growth, and growth is a balance between the rate of production of new individuals and the death rate. A study of these factors is an important branch of ecology, and the study of the growth curves of microorganisms provides a useful model from which to start such investigations.

Until the final phase of decline, *figure 1.14* shows the typical S-shape (sigmoid curve) already seen for individual growth. The curve can be divided into several phases.

■ **Lag phase** Bacteria are adjusting to the conditions. They may, for example, be synthesising new enzymes in order to digest the available nutrients. Growth rate increases slowly.
■ **Exponential phase (logarithmic** or **log phase)** This phase is short. With exponential growth, the numbers double for every unit of time. The resulting growth curve can be converted to a straight line (as in *figure 1.14*) by plotting the logarithms ($\log_{10}$) of the numbers of bacteria on

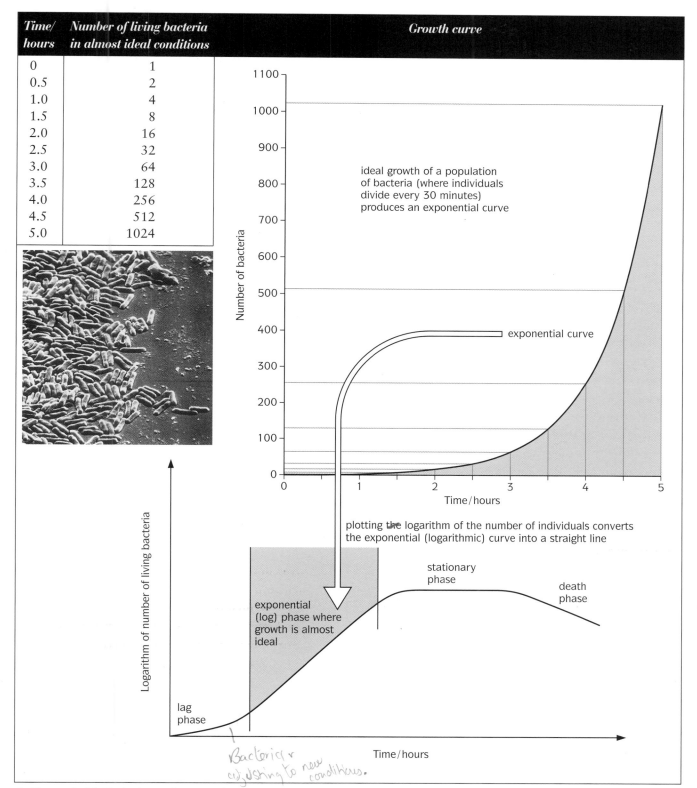

Time/ hours	Number of living bacteria in almost ideal conditions
0	1
0.5	2
1.0	4
1.5	8
2.0	16
2.5	32
3.0	64
3.5	128
4.0	256
4.5	512
5.0	1024

Growth curve

ideal growth of a population of bacteria (where individuals divide every 30 minutes) produces an exponential curve

exponential curve

Number of bacteria

Time/hours

plotting the logarithm of the number of individuals converts the exponential (logarithmic) curve into a straight line

Logarithm of number of living bacteria

stationary phase

death phase

exponential (log) phase where growth is almost ideal

lag phase

Time/hours

Bacteria adjusting to new conditions.

● **Figure 1.14** Typical growth curve of a bacterial population. Growth in numbers of bacteria is shown together with two ways of representing growth graphically. In the top graph the vertical axis is linear; in the bottom graph it is logarithmic. The top graph shows only the exponential (logarithmic) phase of growth in an ideal situation. On the left a scanning electron micrograph of dividing bacteria can be seen.

the vertical axis. This is the most rapid phase of growth, because there are few restrictions. Nutrients and oxygen, for example, are in plentiful supply.

■ **Stationary phase**   Gradually, nutrient and oxygen levels start to decline and waste products build up which may be toxic. Growth rate starts to decline until, in the stationary phase, the rate of death equals the rate of reproduction, and overall population growth is zero.

■ **Death phase** or **phase of decline**   Conditions have deteriorated to the point where death rate exceeds reproduction rate, and total numbers of living bacteria start to decline.

We can learn some important principles from laboratory studies of microorganisms. However, understanding the population growth of organisms under *natural* conditions is often far more difficult. This is because many more factors are important variables. Such factors include competition (between species as well as within a species), disease, climate, predation, population density and parasitism. Their study is part of a branch of biology known as population ecology, but other areas of biology, such as animal behaviour, are also relevant. Observations of natural populations suggest that under a given set of environmental conditions, natural population sizes tend to stabilise and stay reasonably constant over long periods of time. In other words, they typically show sigmoid growth curves and do not enter a phase of decline. Each comes into a dynamic equilibrium with its environment. This was first pointed out by the clergyman and mathematician Thomas Malthus at the end of the eighteenth century and was one of the observations which formed the basis of Darwin's theory of natural selection. Humans are an interesting example of an animal species which is not at present in equilibrium with its environment, and how we achieve such an equilibrium presents us with one of our greatest challenges as a species.

SUMMARY

■ Growth can be defined in a number of ways. The best general definition is a permanent increase in dry mass of living material.

■ Growth is closely linked with development, a progressive series of changes which includes cell specialisation (differentiation) and results in greater complexity.

■ Growth is complex and cannot easily be measured by a single variable. Different methods are used, each with its own particular advantages and disadvantages. For example, increase in dimensions, dry or fresh mass, or cell numbers can be measured. Absolute growth (actual growth), absolute growth rate (change in rate of growth with time) or relative growth rate (which takes into account size) can be plotted as graphs known as growth curves.

■ Different patterns of growth occur. For example, insects show two different types of metamorphosis.

■ Microorganisms provide a useful simple model of population growth.

Questions

1 Discuss the usefulness of unicellular organisms as models for population growth.

2 Explain what is meant by the terms **growth** and **development**.

3 Distinguish between   **a** development and differentiation   **b** individual growth and population growth.

4 Describe, giving full experimental details, how you could measure the absolute increase in dry mass of a plant from seed to maturity.

5 Name an insect showing complete metamorphosis. Outline **four** methods by which you could measure its growth and summarise the advantages and disadvantages of each method.

6 Growth may be measured as absolute growth, absolute growth rate or relative growth rate. Explain the **advantages** of each as a measure of growth.

Asexual reproduction

1 review the range of organisms in which asexual reproduction is found;

2 describe asexual reproduction using one example from each of the five kingdoms: Prokaryotae, Protoctista, Fungi, Plantae and Animalia;

3 discuss the advantages and disadvantages of asexual reproduction as it occurs naturally and explain its evolutionary significance;

4 describe how knowledge of asexual reproduction, growth and development has been used commercially to develop methods of artificial propagation (cloning);

5 discuss the advantages and disadvantages of cloning.

Reproduction is the production of a new organism or organisms by an existing member or members of the same species. No living organism is immortal, so reproduction is essential for the renewal and survival of a species. This is its primary function, but it is also the basis of population growth and spread.

Reproduction takes place in two ways, asexual and sexual. Some organisms use both methods, others just one. **Asexual (non-sexual) reproduction** is the production of new individuals from a single parent without the production of gametes. It is particularly common in plants, simple animals and microorganisms. All the individuals produced by one parent are referred to as a **clone** and are genetically identical. **Sexual reproduction**, on the other hand, involves the fusion of two haploid gametes, usually a male and a female gamete, to form one diploid cell, the zygote. This develops into a new organism. The gametes may come from one individual, or from separate male and female parents. Meiosis must be involved at some stage in the life cycle.

SAQ 2.1

Why must meiosis occur somewhere in the life cycle of an organism which reproduces sexually?

For some organisms, such as humans, sexual reproduction is the only natural form of reproduction, although artificial cloning (a form of asexual reproduction) is a theoretical possibility. Our concept of a species is linked to sexual reproduction because, by definition, a species is a group of organisms which can interbreed by sexual reproduction to produce fertile offspring.

Both types of reproduction are widespread and have their own particular advantages and disadvantages. In this chapter we shall focus on asexual reproduction and its usefulness as a strategy for the long-term survival of a species. We shall also see how widespread a strategy it is, and how humans are increasingly able to make commercial use of the process.

The range of living organisms

It is not possible to understand the biological significance of asexual reproduction without first appreciating something of the huge variety of life on this planet. It is estimated that there are at least five million different species of living organisms, ranging from microscopic, single-celled types to the most complex multicellular plants and animals. The variety would be bewildering if we did not make some attempt to place organisms into groups, in other words to *classify* them. The most widely used classification system divides all organisms into five kingdoms (*figure 2.1*). Each of the five kingdoms contains organisms which reproduce asexually. Examples are shown in *figure 2.2*, and reviewed below.

Prokaryotes (bacteria) and protoctists

All unicellular organisms multiply by a process called **fission** in which a parent cell divides into two or more daughter cells. Before the cell divides, its DNA replicates itself so that each daughter cell is *genetically identical to the parent cell*. Normally

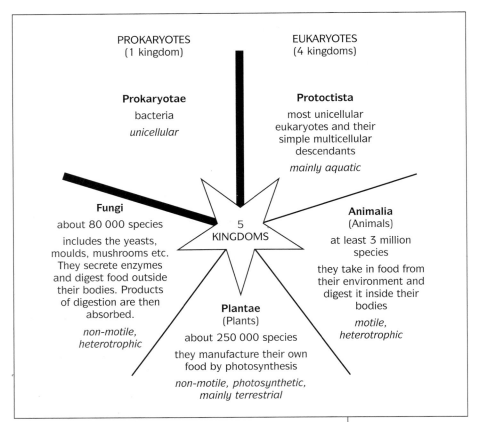

PROKARYOTES
(1 kingdom)

EUKARYOTES
(4 kingdoms)

Prokaryotae
bacteria
unicellular

Protoctista
most unicellular
eukaryotes and their
simple multicellular
descendants
mainly aquatic

5
KINGDOMS

Fungi
about 80 000 species
includes the yeasts,
moulds, mushrooms etc.
They secrete enzymes
and digest food outside
their bodies. Products
of digestion are then
absorbed.
*non-motile,
heterotrophic*

Animalia
(Animals)
at least 3 million
species
they take in food from
their environment and
digest it inside their
bodies
*motile,
heterotrophic*

Plantae
(Plants)
about 250 000 species
they manufacture their own
food by photosynthesis
*non-motile, photosynthetic,
mainly terrestrial*

● *Figure 2.1* The five kingdoms of living organisms. The difference between prokaryotes and eukaryotes is explained in *Foundation Biology* in this series.

the parent cell divides into two, a process called **binary fission**. *Figures 2.2a* and *b* show this process in a bacterium and a protoctist.

Fungi

Most fungi have a characteristic body structure which is made up of a network of fine tube-like threads called **hyphae**. They reproduce asexually by means of **spores**. Spores are formed either directly at the ends of hyphae, as in *Penicillium*, or in special structures called **sporangia**, as, for example, in *Mucor* and *Rhizopus*. *Figure 2.2c* shows asexual reproduction in *Penicillium*, a common fungus. It is a blue-green mould that grows on a wide range of substrates. One species of *Penicillium* is the source of the antibiotic penicillin, and other species are used in the production of some blue cheeses. The spores are just visible to the naked eye. Being light and small, the spores are easily dispersed by air

currents, and germinate to produce new hyphae when they land on a suitable medium. Their large numbers compensate for the high wastage of those that do not find a suitable medium.

Yeasts are unusual fungi in that they are unicellular. They have a form of asexual reproduction known as **budding** *(figure 2.2d)* in which a new individual, identical to the parent, grows from the body of the parent. It starts as a bud and eventually breaks off. Budding also occurs in some plants and animals (see below).

Plants

Asexual reproduction is widespread in the plant kingdom. The most common form is known as **vegetative propagation** and usually involves the growth and development of a bud on part of a stem to form a new plant. This eventually becomes detached from the parent and lives independently. Buds are found only on stems, so the organ of propagation must include at least a small piece of stem, but a wide range of plant structures can be involved. Some common ones are described in *table 2.1*. In many cases the structures serve the equally important function of storing food and surviving adverse conditions such as winter or drought. Stored food can then be used for growth when conditions become suitable.

A convenient organ of vegetative propagation to study is the potato tuber *(figure 2.2e)*. Not only does each potato plant produce more than one tuber, but each tuber can produce more than one plant. Potato tubers in various stages of sprouting can be examined in order to understand the location of buds and the fact that more than one shoot can be produced, thus achieving multiplication as well as reproduction. Artificial cloning of potatoes is discussed later in this chapter.

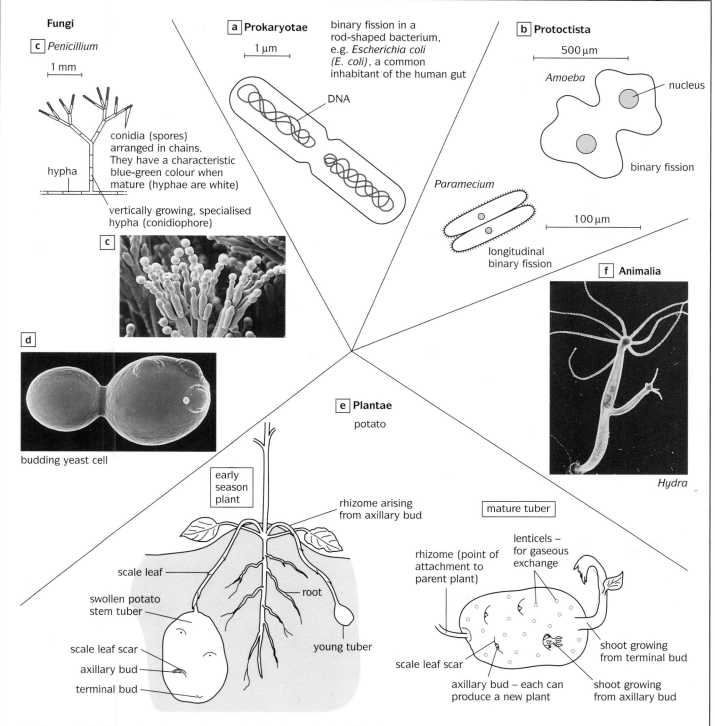

Fungi

c *Penicillium*

1 mm

conidia (spores) arranged in chains. They have a characteristic blue-green colour when mature (hyphae are white)

hypha

vertically growing, specialised hypha (conidiophore)

c

d

budding yeast cell

a **Prokaryotae**

1 μm

binary fission in a rod-shaped bacterium, e.g. *Escherichia coli* (*E. coli*), a common inhabitant of the human gut

DNA

b **Protoctista**

500 μm

Amoeba

nucleus

binary fission

Paramecium

100 μm

longitudinal binary fission

f **Animalia**

Hydra

e **Plantae**

potato

early season plant

rhizome arising from axillary bud

scale leaf

swollen potato stem tuber

root

scale leaf scar

axillary bud

terminal bud

young tuber

mature tuber

rhizome (point of attachment to parent plant)

lenticels – for gaseous exchange

scale leaf scar

axillary bud – each can produce a new plant

shoot growing from terminal bud

shoot growing from axillary bud

● *Figure 2.2* Examples of asexual reproduction from the five kingdoms.

a   Prokaryotae – a bacterium divides into two. Before dividing, the DNA of the cell replicates itself so that each daughter cell is genetically identical to the parent cell. Each daughter cell grows to adult size before dividing again. In the fastest-growing bacteria this may occur every 20 minutes, resulting in rapid population growth (*figure 1.14*).

b   Protoctista – binary fission in *Amoeba* and *Paramecium*.

c   Fungi – diagram and scanning electron micrograph of *Penicillium* showing conidiophores and conidia (spores) (×30).

d   Fungi – scanning electron micrograph of a budding yeast cell used for brewing. Old budding scars are also visible (×5000).

e   Plantae – potato.

f   Animalia – *Hydra* budding (×20) (see also *figure 2.3*).

Name of structure	Plant part	Location	Swollen with food for overwintering?	Notes	Examples
bulb	shoot	underground	yes	stem very short leaves swollen with food	daffodil (*Narcissus*) onion (*Allium*)
corm	stem	underground	yes	stem short	*Crocus, Gladiolus*
rhizome	stem	underground	sometimes, e.g. *Iris*	grows horizontally	mint (*Mentha*), couch grass (*Agropyron*)
stem tuber*	tips of stems or rhizomes	underground	yes		potato (*Solanum tuberosum*)
root tuber*	roots	underground	yes	buds just above root at base of old stem	*Dahlia*
swollen tap root	tap root (main root)	underground	yes	buds just above tap root at base of old stem	carrot (*Daucus*), swede (*Brassica napus*)
stolon	stem	above ground	no	arches over and touches ground. New roots and a new shoot grow from a bud at this point.	blackberry (*Rubus*), gooseberry (*Ribes*)
runner	stem	above ground	no	a type of stolon that elongates rapidly and tends to grow along the surface of the ground. It may be the main stem or grow from one of the lower buds on the main stem	strawberry (*Fragaria*), creeping buttercup (*Ranunculus repens*)

* Unlike bulbs and corms, tubers survive only one year. The following year entirely new tubers are formed.

● **Table 2.1** Organs of vegetative propagation in plants

Animals

Asexual reproduction is confined to animals which have a relatively simple structure. One group of animals, the cnidarians, which includes jellyfish, sea anemones, corals and *Hydra*, shows asexual reproduction in the form of budding. This process has already been described for yeast above, and is shown for *Hydra* in *figures 2.2f* and *2.3*.

Another form of asexual reproduction is **fragmentation**. Here the body breaks into two or more parts, each of which regenerates a new individual.

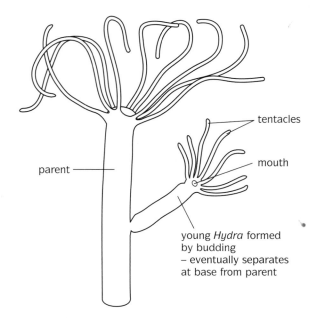

tentacles

parent

mouth

young *Hydra* formed by budding – eventually separates at base from parent

● **Figure 2.3** Budding of *Hydra*, a simple animal.

Ribbon worms, for example, are primitive marine worms whose bodies can break up into pieces, each of which can regenerate.

The artificial cloning of animals is a form of asexual reproduction. The first successful cloning of a vertebrate from a mature adult cell was achieved in the late 1960s when Dr J. Gurdon, who was working at Oxford University, developed a technique for cloning frogs from the skin or intestine cells of an adult frog. This involved removing skin or intestine cells and transplanting their diploid nuclei into the egg cells of another, female, frog. The haploid nuclei of the egg cells were first inactivated by ultraviolet radiation. Thus the new nuclei, with instructions from the donor frog, controlled development of the eggs. The tadpoles, and subsequent frogs, were identical to the donor.

The main application of animal cloning at present is not to produce many identical whole animals, but to maintain identical cells in culture (tissue culture) for a variety of purposes. For example, the effect on cells of new drugs, anti-biotics and other pharmaceutical products, or cosmetics, can be tested without using whole animals. Cell cloning also has important applications in biotechnology. For example, some medically useful proteins, such as growth hormone, are in short supply and can be obtained from cloned mammalian cells. Research is continuing into improved methods of large-scale production. There is more information about this in chapter 6 of *Microbiology and Biotechnology*.

Advantages and disadvantages of natural asexual reproduction

The basis of asexual reproduction is mitosis. This is the division of a nucleus into two identical daughter nuclei. Each has the same genetic make-up because of the replication of DNA in inter-phase. After nuclear division, the rest of the cell divides, thus forming two genetically identical cells. If all the cells so formed remain part of the same organism, it is regarded as growth (see chapter 1). However, if new organisms are formed, asexual reproduction has taken place. It follows that, as stated earlier, all the offspring (clones) produced by one parent as a result of asexual reproduction are genetically identical *(figure 2.4)*.

Prokaryotes do not show mitosis in the same way as eukaryotes, but nevertheless their DNA also replicates before cell division.

Advantages and evolutionary consequences

■ *Only one parent is involved*. This means that, in the case of motile organisms (animals, some prokaryotes and protoctists), time and energy are not used in seeking a mate. For non-motile organisms, the exchange of gametes between separate individuals in sexual reproduction presents problems and special mechanisms are required (see, for example, cross-pollination on page 33). Asexual reproduction avoids these problems.

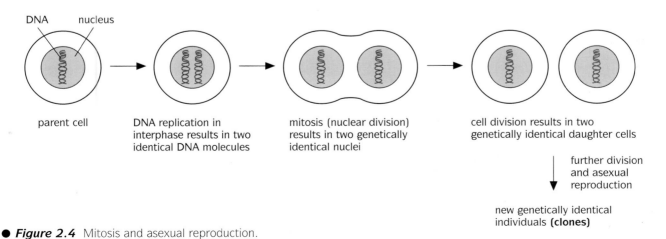

● *Figure 2.4* Mitosis and asexual reproduction.

■ There is *no wastage of gametes*. Production of gametes requires materials and energy and many gametes are inevitably wasted, particularly if released into water as, for example, in the case of many amphibians and fish, mosses and ferns.

■ Asexual reproduction is effective in *dispersing* the species, possibly to exploit suitable habitats some distance from the parent where new resources are available. This is particularly important in fungi since fungi are non-motile, but need to reach new sources of food. In fungi, spores are commonly produced asexually and are ideal for dispersal by air currents since they are light and numerous.

■ Once an organism is established in a particular habitat, it may be able to *spread more effectively* in that habitat (colonise it) by asexual rather than sexual means. For example, grass plants and bracken can spread quite rapidly through an area by means of rhizomes. Cord-grass (*Spartina townsendii*) is a common grass plant that grows on mud flats around our coast (*figure 2.5*). One reason why it is a very effective coloniser of fresh mud is its ability to send out new rhizomes through the mud, each of which can produce a

● *Figure 2.5* Asexual reproduction by rhizomes in *Spartina*, a grass plant which colonises mudflats and salt marshes.

new plant. Sea couch grass (*Agropyron pungens*) does the same on sand dunes.

■ Asexual reproduction may lead to a *rapid production of large numbers of offspring*. New habitats can therefore be exploited rapidly. Microorganisms such as bacteria and fungi provide good examples of this.

■ *Offspring are genetically identical to the parent*. This has important evolutionary consequences because it can be an advantage for the offspring to have the same characteristics as the parent if the latter is well adapted to its environment and is successfully competing with other organisms. Darwin's theory of natural selection states that the 'fitter' members of a species have a greater chance of survival. Asexual reproduction preserves successful combinations of genes.

Disadvantages and evolutionary consequences

■ *If spores are produced, this may be wasteful of materials and energy* since many do not find suitable conditions for germination. However, if such a method of reproduction is a result of evolution, the advantages of producing spores (see above) must outweigh the disadvantages for those species that produce them.

■ Asexual reproduction and consequent spread of an organism in one area can lead to *over-crowding and exhaustion of resources such as nutrients* (see bacterial population growth, page 12).

■ The major disadvantage of asexual reproduction, and probably the reason why sexual reproduction evolved, is that *no genetic variation occurs among the offspring*. As we saw above, there are circumstances in which this is an advantage. However, Darwin's theory states that evolution proceeds by natural selection. This in turn depends on variation existing among the members of a species since the 'fitter' variants are the ones selected when there is some change in the environment. A good example of this was the introduction of the virus which causes myxomatosis into the rabbit population of Britain in 1952. Although very effective at reducing the rabbit population, a few rabbits proved resistant and survived to multiply. After many years, the rabbit population developed a high level of resistance. The virus was originally obtained from Australia, and there the rabbit is now a major pest again, with a population estimated at 2–300 million. In 1994, tests began in Australia with a new virus to try to reduce the population again. This new virus, rabbit haemorrhagic virus, emerged in China in 1984 and has since spread through four continents, killing hundreds of millions of rabbits. Unless there was genetic variation among the rabbits they would be in danger of being wiped out completely. Sexual reproduction helps to increase variation within a population.

Conclusions

In looking at the advantages and disadvantages of asexual reproduction, and comparing asexual with sexual reproduction, we are not trying to judge which is the better form of reproduction. Both can be successful strategies depending on the circumstances. Both have certain advantages. Some organisms use both strategies, others have come to rely exclusively just on one. For example *Amoeba*, a one-celled protoctist, has never been observed to carry out sexual reproduction, and many animals, including humans, do not show asexual reproduction.

Artificial propagation of plants (cloning)

The origins of agriculture go back about 10 000 years, when people in the Middle East began to cultivate cereal crops and changed from a hunter–gatherer existence to a more settled way of life based on farming. Farming demanded a basic understanding of plant growth, development and reproduction. Although our knowledge of these processes has developed enormously since, it is still true that improvements in our understanding of them are likely to lead to the more efficient use of plants for our own needs. Agriculture is now the

world's largest industry in economic terms and current progress is rapid. Revolutionary new methods of propagating plants artificially have been developed over the last 30 years. Also, new food crops, and crops grown for expensive products such as drugs and perfumes, are being developed, and there are continuing efforts to genetically improve existing crops and the efficiency of their production. This is particularly true in the branch of agriculture known as horticulture, which is usually associated with the intensive production of high-value crops such as flowers, vegetables, shrubs and fruit trees.

We shall be looking here at how our knowledge of plant growth and development has been used commercially to develop methods of artificial propagation. A number of such methods exist, and are now considered in turn.

Cuttings

Cuttings are parts of plants removed by cutting which, when placed in suitable conditions, root and produce new plants. Rooting hormone is sometimes added to stimulate rooting. Some plants which are commonly reproduced artificially in this way are: house plants such as *Geranium*, *Pelargonium* and *Coleus*, in which the shoots are used as cuttings; *Forsythia*, a common shrub which produces yellow flowers in spring, in which the shoots (twigs) are used as cuttings; the African violet, another popular house plant, in which the leaves are used as cuttings; the blackcurrant, commercially important for the manufacture of fruit drinks, in which the shoots are used; and *Chrysanthemum*, a popular autumn flower, in which the shoots are used.

Grafting

Grafting is the transplantation of part of one plant, the **scion**, onto the lower part of another, the **stock**. The new plant usually has the root system of the stock and develops the shoot system of the scion. The stock is chosen for its vigour and the shoot is usually chosen for its superior flowers or fruit. This is a common method of propagating those fruit trees, like apple, peach and plum, which cannot easily be grown from cuttings. Most rose bushes are also propagated by this method since they do not breed true from seed. In this case, buds are typically grafted onto woody stocks, such as the wild dog rose. When the scion is a bud rather than a shoot the technique is also known as **budding**. Note that *new* varieties have to be created by *sexual* reproduction.

Layering

Layering is appropriate if the plant produces runners (see *table 2.1*). It is a method traditionally used by gardeners for propagating strawberries and is now also widely used commercially. The technique involves laying out (layering) the runners around the parent plant, pegging them down until they root and then cutting the link between the parent and the daughter plant.

Micropropagation (tissue culture)

'Micro' means small, and micropropagation is the name generally applied to the propagation (cloning) of isolated cells or small pieces of plant tissue in special solutions called culture solutions. The growth of tissue under controlled conditions in such solutions is called **tissue culture**.

From the beginning of the 20th century it was thought that mature plant cells probably had the potential to grow into new plants given the right conditions, but it was not achieved until the 1960s. Then F.C. Steward of Cornell University in the USA grew whole carrot plants from a few mature cells of carrot root, using a culture solution containing a special mix of nutrients and plant hormones. An outline of the techniques used in tissue culture is given below and *figure 2.6* summarises the three basic methods of cloning from a stock plant. *Figures 2.7* to *2.11* show some of the typical stages in the production of cloned plants.

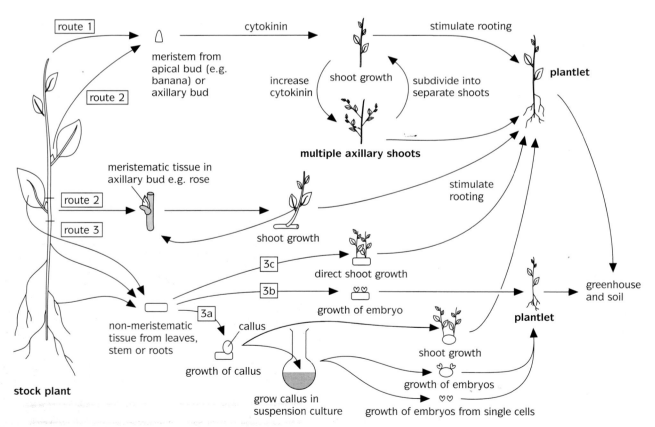

● *Figure 2.6* Methods for cloning from a stock plant. Three routes are shown. Routes 1 and 2 involve the use of meristematic tissue (tissue with dividing cells) either from apical buds (route 1) or from axillary buds (route 2). Route 3 involves the use of non-meristematic tissue. This can sometimes produce new embryos as in routes 3a and 3b.

● *Figure 2.7* Cutting up cultured tissue for planting as individual plantlets.

The culture medium

A typical culture medium contains sucrose as a source of energy, other organic nutrients such as amino acids and vitamins, and a wide range of inorganic ions such as nitrate, potassium, phosphate and trace amounts of iron and copper. It also contains a balance of growth regulators (hormones), particularly **auxins**, which stimulate root growth, and **cytokinins**, which stimulate shoot growth.

Unfortunately, the culture medium also provides ideal growth conditions for bacteria and fungi, which grow so fast that they can overwhelm the plant tissue. For this reason all procedures must be carried out under aseptic (sterile) conditions (*figures 2.7* and *2.8*). This involves disinfecting the original plant tissue and preventing any subsequent contamination by growing it in special transparent containers. These are placed in growth rooms under controlled environmental conditions.

● *Figure 2.8* Transferring cultured plantlets to growing medium under aseptic conditions.

Mass production

Once the growing tissues have produced new shoots, they can be subdivided to produce new plants. This is routinely done at regular intervals, typically every four to eight weeks (*figure 2.7*). In practice, the number of plants can sometimes be increased two- to four-fold every four to eight weeks, and over a year thousands or even millions of identical plants can be created from one original culture. The young 'plantlets' can be kept in cold storage to build up numbers before growing into mature plants. This is necessary if a bulk supply is required, for example for a new plantation of trees or the planting out of a crop at the right time of year.

One or more **stock** plants must first be produced with the desired properties for cloning. Small pieces (**explants**) are taken from these as shown in *figure 2.6*. This figure also shows three basic approaches to cloning from the stock plant (routes 1, 2 and 3). The most important route commercially is to start with meristematic tissue and develop many axillary shoots. The other routes have more specialised uses.

Another method shown in *figure 2.6* is to produce a callus from non-meristematic (non-dividing) plant cells (route 3a). A **callus** is a mass of disorganised cells. The callus can be induced to form roots or shoots by carefully balancing the amounts of auxin and cytokinin in the medium (*figure 2.9*). Given the right conditions, cells from callus tissue or from non-meristematic tissue sometimes behave as if they are zygotes and develop

● *Figure 2.9* Young shoots arising from callus tissue derived from an anther.

into embryos and then into seedlings *(figures 2.6 (routes 3a and b) and 2.9)*. Embryos can be embedded in pellets of alginate jelly to produce **artificial seed**.

Table 2.2 shows that in 1988, the Netherlands, Europe's largest producer of micropropagated plants, produced over 60 million plants by these methods. The majority were pot plants, cut flowers and ornamental corms and bulbs, such as tulips. Over two million orchids were produced. Large-scale production of agricultural crops is less common, with the exception of potatoes, where micropropagation is used to produce virus-free plants. The current technique is to produce potato plantlets, which are formed in large numbers from axillary shoot cultures as shown in *figure 2.6*. These then produce minitubers which can be sown directly in fields and generate normal plants. Over half a million minitubers can be produced from one original tuber in one year using micropropagation. At the end of the second year, the potato yield from these is over three million kilograms of tubers.

Britain has built up a sizeable export market in tropical and subtropical plantation crops such as banana, sugar cane, date and oil palms. Oil palms, in particular, have been produced in large numbers from callus tissues, allowing production of unlimited numbers of stable, uniform types. *Figure 2.10* shows the mass production of orchids, in this case in Thailand.

Type of product	Number of plants propagated
Ornamental plants	
pot plants, e.g. cyclamen	26 730 000
cut flowers	18 231 000
ornamental bulbs and corms	12 951 000
orchids	2 448 000
plants for aquaria	412 000
carnivorous plants	10 000
Trees and shrubs – ornamental and fruit	193 000
Agricultural crops	
potato	395 000
sugar beet	37 000
rye grass	2 000
vegetables	93 000
Total	61 502 000

● *Table 2.2* Production of micropropagated plants in the Netherlands in 1988

● *Figure 2.10* Commercial micropropagation. Rows of orchid cultures, each in a sterile atmosphere, in Bangkok, Thailand.

● *Figure 2.11* Plants in a fogging greenhouse – the fog is a cloud-like suspension of very fine water droplets providing ideal conditions for growth.

Advantages of artificial propagation

The chief advantage of artificial propagation is that the cloned plants are genetically identical to the parent plant. Many copies of plants with desirable characteristics can therefore be produced. The selective breeding of plants, involving sexual reproduction, is probably as old as agriculture itself. However, it is difficult to produce true breeding lines when relying on sexual reproduction, particularly when a plant is adapted for cross-pollination, such as the apple tree. Many generations of selection may be necessary. By cloning, particular combinations of genes can be fixed in just one step. Buyers can be supplied with a standard product.

In some cases, artificial propagation is simply a more convenient method of propagating plants than sowing seeds. Taking cuttings of blackcurrant bushes or layering strawberry runners, for example, are rapid and simple procedures, and cut out the need to nurture delicate seedlings.

Some plants, such as the commercially grown banana, are sterile or suffer from low germination rates. In others, it may be difficult to set up the correct conditions for successful seed germination, as with orchids. In these cases, cloning may be a more practical alternative.

Tissue culture has particular benefits associated with it.

- Rapid production of large numbers of plants from just one or a few stock plants can be achieved. Use of the technique is increasing every year.
- Plant diseases can be avoided. For example, virus-free plants can be obtained by selecting only meristematic tissue. Viruses tend to be distributed throughout the plant body by the vascular system, but the meristems lack vascular tissue and are therefore free of most viruses. They are usually also bacteria-free. The meristem can subsequently be heat-treated to kill most or all remaining viruses and bacteria since they are usually more sensitive to heat than the meristematic cells. As already mentioned, micropropagation is important in the production of virus-free potatoes.

- Micropropagated plants can be produced at any time of year and can also be put in cold storage, taking up relatively little space. This gives great flexibility in supplying consumer demand. Out-of-season plants sell for higher prices. Combined with rapid production, this opens up new marketing possibilities. It is envisaged, for example, that it will be possible to produce new varieties of house plants and garden ornamentals more quickly. Then, just as other commodities such as clothes and cars are subject to fashion, so we will tend to buy the latest plants to arrive in the supermarket or garden centre. This, in turn, will increase commercial turnover and profit.
- Many identical plants can be produced for subsequent plant breeding programmes that require sexual reproduction.
- A market can be created for exotic plants such as orchids and insectivorous plants that are hard to produce in large quantities from seed.
- Micropropagation can be linked with genetic engineering. If a new gene is introduced into a plant cell by genetic engineering, the modified cell can be grown into a whole plant and then cloned to produce many new plants, all containing that gene.
- To some extent, plants can be designed to order. The ornamental plant *Ficus*, for example, can be produced in a single-stemmed or a multi-stemmed form.
- Their light weight and small size mean that micropropagated plants can be airfreighted easily and cheaply, thus increasing international trade.
- Standardising the growing conditions produces batch after batch of standard plants. Such reliability is an important sales advantage.

SAQ 2.2

A garlic bulb produces a single plant which in turn produces a group of new bulbs at the end of the growing season. Suggest **a two** advantages and **b two** disadvantages of growing garlic from bulbs rather than from seeds.

Disadvantages of artificial propagation

Artificial propagation is not as convenient as sowing seed. When very large numbers of plants are required, it is often impractical. It would be desirable, for example, to produce certain vegetables such as carrots and celery by artificial propagation because these plants normally show cross-pollination, which can lead to undesirable variation. However, the planting out of sufficient identical seedlings would be too time consuming and uneconomic.

All the clones are genetically identical. Therefore any change in environmental conditions, or the appearance of a new disease, could have devastating consequences if the plants are not resistant or cannot adapt.

Tissue culture, in particular, has certain disadvantages associated with it:

■ The processes are labour-intensive and there are therefore high labour costs. Also, the equipment needed is expensive. In the future, it is likely that more automation will be introduced. Because of the cost, the individual plants produced must have a high market value and this is why tissue culture is mainly confined to ornamental rather than crop plants *(table 2.2)*.
■ The work must be carried out in sterile conditions. This requires highly trained staff and imposes severe constraints on working practice.
■ Since the techniques are relatively new, unforeseen problems have sometimes arisen. For example, it was decided in the 1970s to replace oil palms on Malaysian plantations with new micropropagated varieties. Five years later, when the first fruit (from which the palm oil is extracted) should have been produced, they were discovered to be sterile. The problem was traced to genetic changes that had taken place in culture. Such genetic changes seem to be more common in tissue culture and strict quality controls are needed to avoid such problems.

SUMMARY

■ Asexual reproduction is the production of new individuals from a single parent without the involvement of meiosis and gametes. Its basis is mitosis.

■ Asexual reproduction occurs in all the five kingdoms of living organisms.

■ Methods of asexual reproduction vary and include fission (for example bacteria and some protoctists), spore production (for example many fungi), budding (for example *Hydra* and yeast), and various methods of vegetative propagation in plants.

■ There are advantages in reproducing asexually. These include the need for only one parent and no wastage of gametes. In some species, asexually produced spores are used for dispersal. In some, rapid multiplication and spread is achieved asexually. Production of genetically identical offspring is an advantage where an individual is well adapted to its environment.

■ Disadvantages of asexual reproduction include the possibility of overcrowding in an area and the lack of genetic variation among offspring. Genetic variation increases the survival chances of a *species* because it is more likely that some individuals will survive adverse conditions.

■ Our knowledge of asexual plant propagation (cloning) has led to commercial exploitation of the process. Various methods are used, including cuttings, grafting, layering and, more recently, tissue culture (micropropagation).

■ Tissue culture is of great commercial importance. It is used particularly for producing flowers, ornamental plants and virus-free potatoes.

■ Cloning allows mass production of identical, disease-free plants selected for their ideal characteristics. Plants can be put into long-term storage and produced at any time of year. More flexibility in creating new varieties and meeting consumer demands is possible.

■ Cloning of some plants is not practicable. Tissue culture is labour-intensive, relatively expensive and requires sterile operating conditions. Some unexpected problems arise from time to time because techniques are new. Cloned plants are genetically identical, so all plants are equally vulnerable to new diseases or environmental change.

Questions

1 **a** Discuss the commercial **advantages** of cloning plants.

 b Discuss the possible **disadvantages** associated with cloning.

2 Define **a** asexual reproduction **b** sexual reproduction. Discuss the **advantages** of both types of reproduction.

3 Review, giving named examples, the range of methods used by animals and plants to carry out asexual reproduction under natural conditions.

4 Explain how mass production of plants can be achieved by artificial means.

5 Explain the need for the following in work with tissue cultures:

 a growth regulators **b** aseptic technique.

Chapter three

Sexual reproduction in flowering plants

1 recognise and name the main parts of a typical flower;

2 describe anther structure and pollen formation;

3 describe the development of the ovule;

4 distinguish between self-pollination and cross-pollination;

5 describe and explain the features of a typical insect-pollinated and a typical wind-pollinated flower;

6 explain the relative merits of self-pollination and cross-pollination;

7 describe mechanisms favouring cross-pollination and self-pollination;

8 describe double fertilisation in the embryo sac, and explain its significance;

9 carry out an experiment to observe pollen tube growth;

10 describe development of the ovule into the seed, including embryo development;

11 describe development of the ovary into the fruit;

12 carry out an experiment to investigate embryo development in shepherd's purse;

13 describe epigeal and hypogeal germination.

Sexual reproduction is the fusion of two gametes to form a diploid cell, the **zygote**, that develops into a new organism. Of particular importance is the fusion of the nuclei of the gametes because this brings two separate sets of chromo-somes together. In flowering plants, male and female gametes may be produced in the same plant or in separate plants. They are produced in special structures, the **flowers**, which are unique to this group of plants. *Figure 3.1* shows an outline of the life cycle of a flowering plant. In this chapter we shall be looking at the different stages of the life cycle, so *figure 3.1* should be referred to from time to time to keep an overview of the whole process. *Figure 3.1* shows that male gametes are made inside **pollen grains**, and female gametes inside structures known as **embryo sacs**.

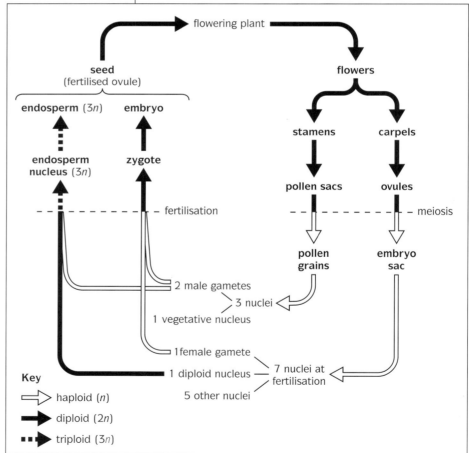

● *Figure 3.1* Life cycle of a flowering plant

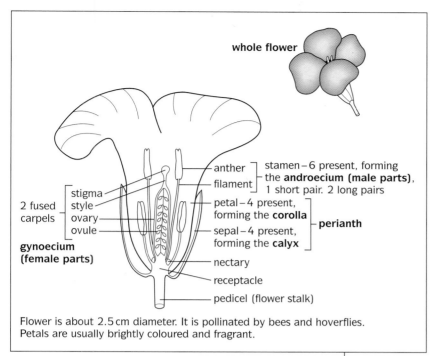

whole flower

anther
filament
stamen – 6 present, forming the **androecium (male parts)**, 1 short pair. 2 long pairs

stigma
style
ovary
ovule

2 fused carpels

gynoecium (female parts)

petal – 4 present, forming the **corolla**
sepal – 4 present, forming the **calyx**
} **perianth**

nectary
receptacle
pedicel (flower stalk)

Flower is about 2.5 cm diameter. It is pollinated by bees and hoverflies. Petals are usually brightly coloured and fragrant.

● *Figure 3.2* Half-flower of wallflower. The female parts of the plant (carpels) are known collectively as the gynoecium; the male parts (stamens) as the androecium.

The parts of a flower

It is very helpful to be familiar with the main parts of a flower before trying to understand the life cycle. The wallflower (*Cheiranthus cheiri*) is a good example to study because it has a simple flower which is available relatively early in the year (March to June). *Figure 3.2* shows the structure of its flower.

In general, a flower has a stalk called the **pedicel**, the top end of which is swollen to form the **receptacle**. The **petals** and **sepals** grow from the receptacle. The ovary is found just *above* the receptacle (**superior ovary**) or just *below* (**inferior ovary**). It is made of one or more **carpels**. These may be separate (free) as in the buttercup, or more commonly fused to make a single ovary, as in the wallflower. Carpels contain **ovules**, the future

seeds, and are the equivalent of **stamens** where the **pollen** develops. A group of flowers on one stalk is called an **inflorescence**.

Development of pollen grains

Pollen grains are formed inside the anthers in structures called **pollen sacs** (*figures 3.3* and *3.4a*). Within the pollen sacs are many pollen mother cells which, like the other cells of the flowering plant, have diploid nuclei (two sets of chromosomes). Each mother cell nucleus divides by meiosis to form four haploid nuclei, each with one set of chromosomes. The nuclei are within daughter cells whose walls thicken to form pollen grains (*figure 3.4b*). At first the four cells are grouped together, forming a **tetrad**. Later, they separate and develop into pollen grains. Each pollen grain has a thick, sculptured wall, the pattern being characteristic of the species or genus. The outer wall, or **exine**, is made of an extremely resistant chemical called sporopollenin which can enable pollen grains to survive for long periods, in some cases for millions of years.

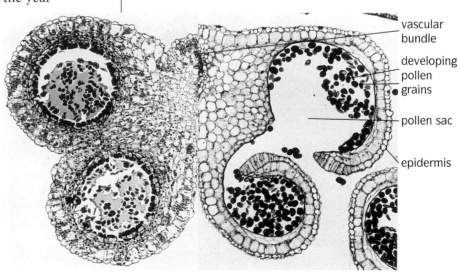

vascular bundle
developing pollen grains
pollen sac
epidermis

● *Figure 3.3* Photomicrograph of TS anther of *Lilium* before dehiscence (left-hand side) and after dehiscence (right-hand side).

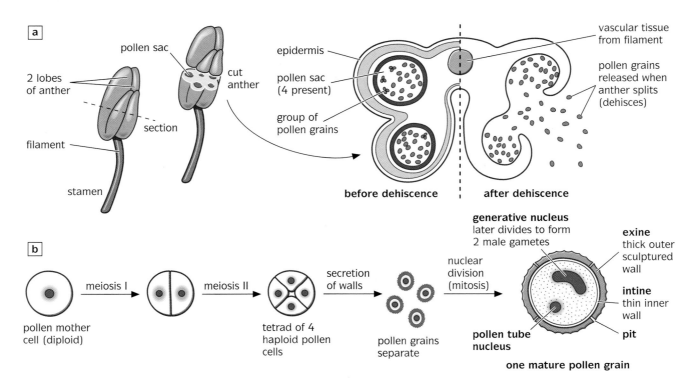

a 2 lobes of anther
   pollen sac
   cut anther
   section
   filament
   stamen

   epidermis
   pollen sac (4 present)
   group of pollen grains

   vascular tissue from filament
   pollen grains released when anther splits (dehisces)

   before dehiscence   **after dehiscence**

b pollen mother cell (diploid)
   → meiosis I →
   → meiosis II →
   tetrad of 4 haploid pollen cells
   → secretion of walls →
   pollen grains separate
   → nuclear division (mitosis) →

   generative nucleus
   later divides to form 2 male gametes

   exine thick outer sculptured wall
   intine thin inner wall
   pollen tube nucleus
   pit

   one mature pollen grain

● *Figure 3.4*
a TS anther before and after dehiscence.
b Development of pollen grains.

The pollen grain nucleus divides into two by mitosis forming a **generative nucleus**, which later divides to form two male gametes, and a **pollen tube nucleus** *(figure 3.4b)*. The structure of the anther and development of the pollen can be observed in sections of mature anthers *(figure 3.3)*. When the pollen is mature, the anthers dry and split open (**dehisce**) and release the pollen.

Development of the ovule

This is best summarised by means of diagrams *(figure 3.6)*. During development of the female gamete, the nucleus of the embryo sac mother cell divides by meiosis to form four haploid nuclei within four cells. Three of these cells degenerate. One becomes the **embryo sac**. The embryo sac nucleus divides three times by mitosis to form eight nuclei, four at each end of the embryo sac. One nucleus from each end migrates to the centre of the embryo sac and these two nuclei fuse to form a diploid nucleus which will later fuse with the nucleus of one of the male gametes. The remaining

six nuclei, three at each end, become separated by thin cell walls. Of the six cells formed, one becomes the female gamete, or ovum, which later fuses with the second male gamete. The others appear to have no function and soon disintegrate. *Figure 3.5* shows an embryo sac at fertilisation.

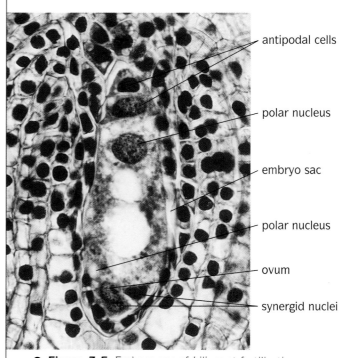

antipodal cells
polar nucleus
embryo sac
polar nucleus
ovum
synergid nuclei

● *Figure 3.5* Embryo sac of *Lilium* at fertilisation.

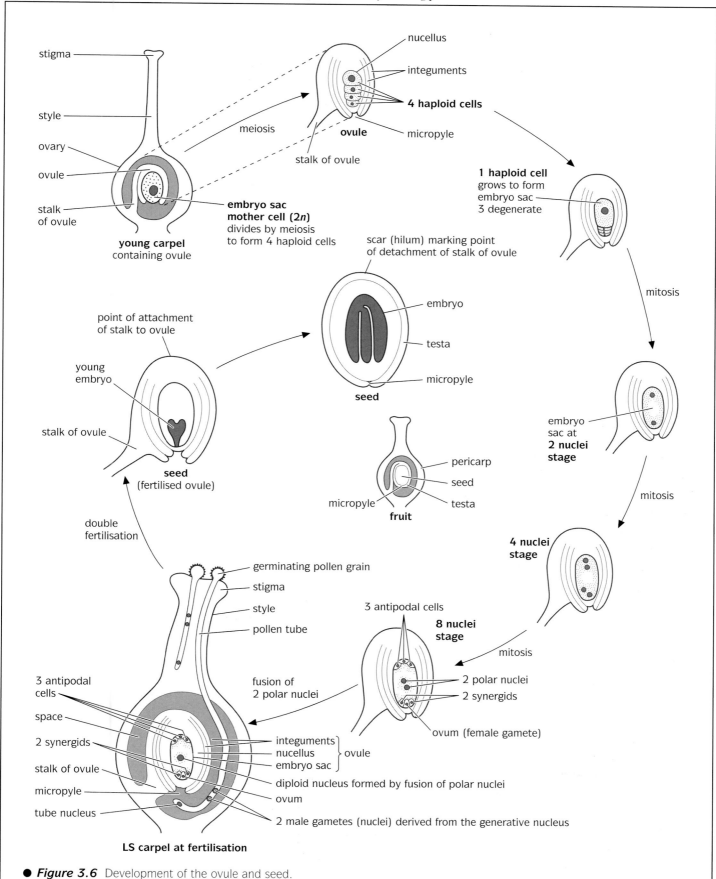

● **Figure 3.6** Development of the ovule and seed.

Pollination

Pollination is the transfer of pollen grains from the anther to the stigma. **Self-pollination** is transfer to a stigma on the same flower *or* to a different flower on the same plant. **Cross-pollination** is transfer to a stigma on another plant. *Do not confuse pollination with fertilisation.*

Pollination is necessary in order to bring the two male gametes, which are inside the pollen grain, to within close proximity of the female gamete. The male gametes are protected from drying out during this transfer by the wall of the pollen grain.

When anthers are mature, they dry and split (**dehisce**) down their lengths along two lines of weakness, thus releasing the pollen grains. A number of mechanisms have evolved to help to ensure successful cross-pollination, the two most common being **wind pollination** and **insect pollination**. Flowers are often highly adapted for one of these mechanisms. Two examples will be examined, namely white deadnettle (*Lamium album*) for insect pollination, and a grass (meadow fescue, *Festuca pratensis*) for wind pollination. Other suitable examples of insect-pollinated flowers are the bluebell (which flowers from April to June), the hyacinth (flowering in May) or the sweet pea (flowering in July). Another wind-pollinated grass which could be studied is the cereal oat (*Avena sativa*), which has relatively large flower parts.

Insect pollination

Figure 3.7 shows the structure of the white deadnettle flower and indicates some of the ways in which the flower is adapted for insect pollination. It is pollinated by long-tongued insects, such as bumble bees, which can reach the nectaries from the landing platform formed by the lower lip of the flower. The nectaries secrete nectar, a liquid rich in sugars, amino acids and other nutrients which the bees feed on. They may also collect pollen for food. The stigma projects below the anthers, so as the bee enters the flower its back touches the stigma first. Its back may be carrying pollen from a previous visit to another deadnettle flower. Pollination occurs when pollen from the back of the bee is transferred to the stigma. This will usually be cross-pollination, although self-pollination can also occur. Then the anthers touch the bee's back, shedding pollen onto the bee. To make cross-pollination more likely, the anthers mature before the stigma. The cluster of flowers, the inflorescence, is conspicuous, helping to attract the bees. The relatively large white flowers are also conspicuous amongst green vegetation.

Wind pollination

Meadow fescue has flowers typical of grasses in being very small and inconspicuous, since they have not evolved to attract insects. Instead, pollination is

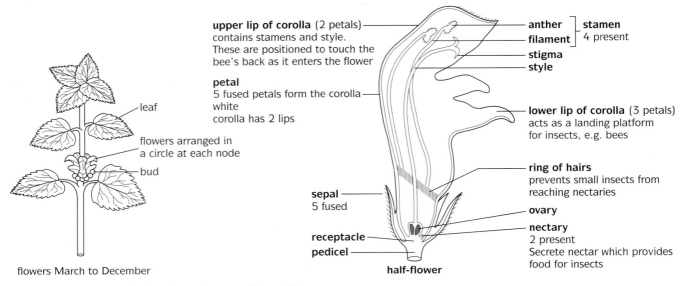

● **Figure 3.7** White deadnettle, an insect-pollinated flower.

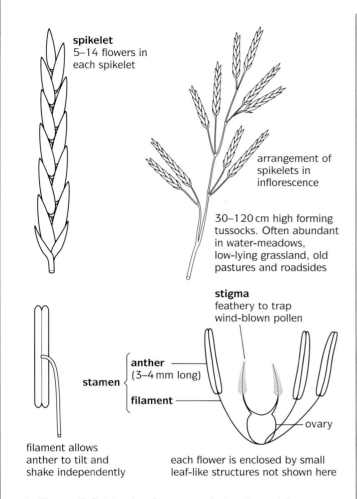

spikelet
5–14 flowers in
each spikelet

arrangement of
spikelets in
inflorescence

30–120 cm high forming
tussocks. Often abundant
in water-meadows,
low-lying grassland, old
pastures and roadsides

stigma
feathery to trap
wind-blown pollen

anther
(3–4 mm long)

stamen

filament

ovary

filament allows
anther to tilt and
shake independently

each flower is enclosed by small
leaf-like structures not shown here

● *Figure 3.8* Meadow fescue, a wind-pollinated flower.

by means of the wind, which is more wasteful of
pollen than insect pollination. It helps, however,
that grass plants, like most wind-pollinated plants,
tend to live close to one another. *Figure 3.8* shows
the structure of the meadow fescue flower, which is
similar to that of most grasses. There are no petals,
nectaries or scent to attract insects. However, the
stigma is relatively large and feathery and is an
effective pollen trap. The stamens hang outside the
structures when ripe and can swing freely in air
currents. They produce large quantities of small,
light pollen grains with smooth surfaces which are
easily dispersed by air currents. The flowers are
borne on tall, loose, nodding inflorescences which
also catch the wind easily.

When examining grass flowers, a hand lens or
dissecting microscope and one or two dissecting
needles are recommended since the flowers are

very small and have to be carefully removed from
surrounding leaf-like structures. Most grasses
flower during May, June and July.

SAQ 3.1

Make a table to compare the adaptations for pollination
shown by flowers of the white deadnettle (insect pollinated) and meadow fescue (wind pollinated).

The relative merits of self- and cross-pollination

There are advantages associated with both self- and
cross-pollination. In fact, many plant species show
both types. Self-pollination is a more reliable form
of pollination, particularly if members of the species
are widely scattered. However, self-pollination
results in self-fertilisation, where gametes produced
by the same parent fuse. This results in inbreeding
and can have genetic disadvantages. For example,
harmful recessive characteristics are more likely to
be expressed. On the other hand, where a plant is
well adapted for a constant environment, genetic
uniformity may be an advantage (see page 20). Self-
pollination is also advantageous in harsh environ-
ments, such as high on mountains, where insects
and other pollinators may be scarce.

SAQ 3.2

Explain why **a** self-fertilisation can still result in genetic
variation among the offspring **b** it makes no difference
genetically whether self-pollination takes place in the same
flower that produced the pollen or in another flower on the
same plant.

Cross-pollination is less reliable, and more wasteful
of pollen, but genetically it has the advantage of
producing greater variation (outbreeding) because
different plants exchange genes.

Mechanisms favouring cross-pollination

Elaborate mechanisms often exist to increase the
likelihood and efficiency of cross-pollination.

■ **Dioecious plants.** When a species, such as
willow, produces separate male and female

plants it is described as **dioecious**. Self-pollination is impossible in such species, but the number of dioecious species is very few.

■ **Monoecious plants. Monoecious** species, such as oak and birch, are those which produce separate male and female flowers on the same plant. This encourages cross-pollination between adjacent trees, while still allowing self-pollination among the flowers of the same tree.

■ **Protandry and protogyny.** Anthers and stigmas sometimes mature at different times, thus encouraging cross-pollination. If the anthers mature first, as in the white deadnettle, it is known as **protandry**. The term used when the stigmas mature first, as in the bluebell, is **protogyny**. Usually there is an overlap period when both anthers and stigmas are ripe, allowing for self-pollination as well.

■ **Self-incompatibility.** Even if self-pollination occurs, self-*fertilisation* is often made less likely or impossible by slow or zero growth of the pollen tubes. This is genetically determined. Extreme examples are the various species of clover, which are totally self-incompatible.

■ **Special structures.** *Figure 3.9* shows a special mechanism which favours cross-pollination and operates in the primrose (*Primula vulgaris*) (see also *SAQ 3.3*). In this species, self-incompatibility also occurs. Many other, often unique, mechanisms exist.

SAQ 3.3

Examine *figure 3.9* which shows pin-eyed and thrum-eyed flowers of primrose. These are found on separate plants. **a** In one type of flower, the stigma is *above* the anthers and in one type *below*. Which is which?   **b** Suggest how this favours cross-pollination by bees.

Mechanisms of self-pollination

Self-pollination is very common and can probably occur in more than half of all flowering plant species. Where it occurs, flowers are usually **hermaphrodite** (meaning that they have both male and female parts). A simple mechanism of self-pollination is for anthers and stigmas to mature at the same time and for the pollen to be shed directly onto the stigma. Flowers adapted for self-pollination are usually small, inconspicuous and produce no nectar or scent since they do not need to attract insects. Examples are groundsel and chickweed, both common weeds. In some species or individual flowers, cross-pollination is prevented because the flower buds never open.

Fertilisation

If a pollen grain lands on a stigma of a compatible species, it will germinate. A pollen tube emerges from one of the pores (pits) in the pollen grain wall *(figures 3.4b and 3.6)*, and, responding to chemicals secreted by the ovary, grows rapidly down the style to the ovary (an example of **chemotropism**). Growth is controlled by the haploid **tube nucleus** which is found at the tip of the pollen tube.

During growth the haploid **generative nucleus** of the pollen grain divides by mitosis into two haploid nuclei which are the male gametes. The pollen tube enters the ovule through the micropyle, the tube nucleus degenerates and the tip of the tube bursts, releasing the two male gametes. One fuses with the ovum (the female gamete) to form the diploid **zygote**. The other fuses with the diploid nucleus at the centre of the embryo sac to form a **triploid** nucleus, that is, a nucleus

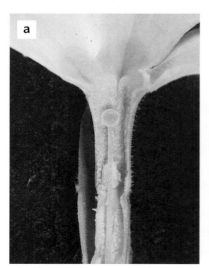
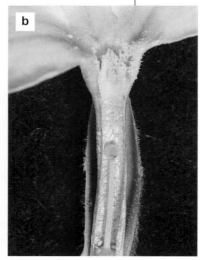

● **Figure 3.9** Cross-pollination in primrose. Two types of flower occur **a** pin-eyed and **b** thrum-eyed.

with three sets of chromosomes. It is known as the **endosperm nucleus**. Thus a **double fertilisation** takes place, a process unique to flowering plants. The zygote will later form the embryo, which will grow into the next generation of the plant, whereas the endosperm nucleus will often form a food store in the seed (see Development of the seed below).

SAQ 3.4

Summarise the roles played by the two pollen grain nuclei.

SAQ 3.5

The male gametes of simple plants such as mosses and ferns are swimming sperm. Suggest why plants evolved a mechanism for carrying male gametes in pollen grains.

It is possible to investigate the germination of pollen grains using the dehiscing anthers of flowers such as *Pelargonium*, wallflower, *Impatiens* or white deadnettle. The pollen grains need to be suspended in a 10% sucrose solution (to stimulate germination) in the central depression of a cavity slide. The sucrose solution should contain borate at a concentration of 0.01% (to stimulate growth and help to prevent osmotic bursting of the pollen tube tips). A drop of aceto-carmine or neutral red can be used to stain the nuclei at the tip of the growing pollen tubes. Once the pollen grains are growing, a microscope with a calibrated eye-piece can be used to measure the lengths of several of the tubes at appropriate intervals (e.g. every three minutes) and thus the rate of growth. The effects of sucrose concentration on the growth rate can also be studied, e.g. by measuring how the rate of growth changes when it is increased from 10 to 20%.

Development of the seed

Immediately after fertilisation the ovule is called the **seed**. Thus a seed is a fertilised ovule. The following changes take place (refer to *figure 3.6* and *figure 3.10* when studying these):

1 The integuments become the **testa**. This is a thin, tough, protective layer around the seed.
2 The **nucellus** disintegrates as the seed develops, supplying nutrients for growth of the embryo and the endosperm in endospermous seeds.
3 The triploid **endosperm nucleus** divides repeatedly by mitosis to form the triploid **endosperm**. The nuclei become separated from one another by thin cell walls. In some seeds (**endospermous seeds**) such as cereals, this remains as the food store for the seed. In **non-endospermous seeds** such as the pea and shepherd's purse, the cotyledons (see below) grow at the expense of the endosperm, which may then disappear altogether.
4 The **zygote** grows by repeated mitotic divisions to become an **embryo**. This consists of a **plumule** (the first shoot), a **radicle** (the first root) and either one or two **cotyledons** (seed leaves) (*figures 3.11* and *3.12*). **Monocotyledons** have

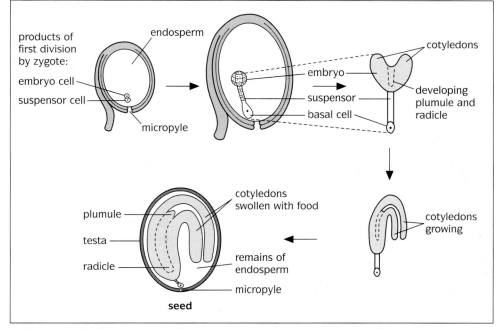

● *Figure 3.10* Development of the embryo in a non-endospermous dicotyledonous seed such as shepherd's purse (*Capsella bursa-pastoris*).

one cotyledon and **dicotyledons** have two. In some seeds, for example pea and broad bean, the cotyledons become swollen with nutrients. The nutrients required may be supplied by the endosperm or the nucellus, or from the parent plant by way of the vascular tissue in the seed stalk.

5　The **micropyle** remains as a tiny pore in the testa through which, later, oxygen and water can reach the seed at germination.

6　As the seed matures, the water content drops markedly from about 90% by mass to about 10–15% by mass. This is in preparation for seed dormancy when metabolic activity will be much reduced. Dry seeds respire extremely slowly and can survive extended drought or cold periods.

7　Remaining flower parts (such as the petals and sepals) wither and die and are shed.

SAQ 3.6

Suggest how the structure of cells may change significantly during preparation for dormancy.

SAQ 3.7

Triploid cells are able to divide by mitosis, but rarely achieve successful meiotic division. Explain why this is so and explain its relevance to the endosperm nucleus.

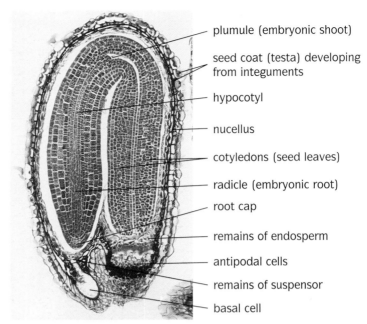

● *Figure 3.11* LS well-developed embryo of shepherd's purse (*Capsella bursa-pastoris*).

Labels: plumule (embryonic shoot); seed coat (testa) developing from integuments; hypocotyl; nucellus; cotyledons (seed leaves); radicle (embryonic root); root cap; remains of endosperm; antipodal cells; remains of suspensor; basal cell

Development of the fruit

While the seed, or seeds, are developing, other changes take place which result in the development of the **fruit**. Just as the ovule becomes the seed immediately after fertilisation, so the ovary becomes the fruit. Thus a fruit is a fertilised ovary. The fruit wall is known as the **pericarp**. The fruit contains the seeds and the pericarp is commonly modified to aid their dispersal by becoming fleshy, or hard and dry. *Figure 3.12* summarises some of the changes associated with fertilisation.

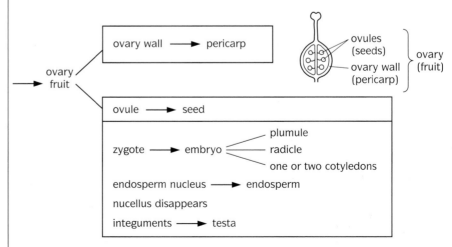

● *Figure 3.12* Summary of the changes in the ovary after fertilisation.

Embryo development can be studied in a common plant of waste ground and gardens, shepherd's purse (*Capsella bursa-pastoris*), so named because the heart-shaped fruits resemble the purses that were once used by shepherds. The best plants to use are those with small white flowers at the apex and a range of sizes of fruits (the youngest are nearest the apex). The ovules can be dissected out of one of the fruits under a microscope or tripod lens and placed in 5% sodium or potassium hydroxide solution for a few minutes. Individual ovules can then be transferred into two drops of 5% glycerine on a clean microscope slide and covered with a cover-slip. When the coverslip is tapped with the handle of a mounted needle, the ovule bursts, and the embryo is exposed. By extracting embryos from a range of fruit sizes, the various stages of embryo development can be followed.

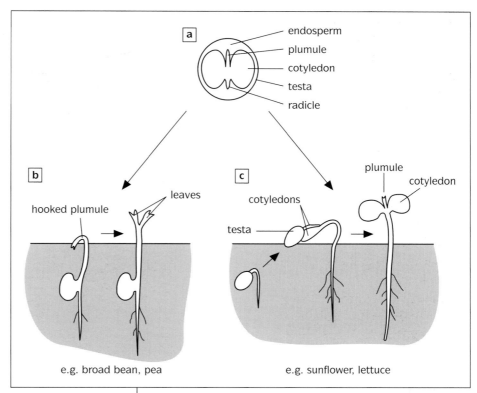

● *Figure 3.13* **a** Simplified diagram of a seed
b hypogeal germination **c** epigeal germination.

Germination

In dicotyledons there are two types of germination, **hypogeal germination** where the cotyledons remain in the seed below ground, and **epigeal germination** where they are carried above ground (*figure 3.13*). In hypogeal germination the first shoot, the plumule, grows from a region just above the cotyledons. The plumule remains hooked to protect the growing tip as it grows through the soil. In the case of epigeal germination, growth occurs from the region just below the cotyledons. Thus the cotyledons, still enclosed in the testa, are carried up through the soil and become the first photosynthetic structures above the soil.

In both cases, as the hook emerges from the soil, a phytochrome-controlled response to light (see chapter 5) results in the straightening of the hook,

and the greening and expansion of the cotyledons or first leaves. Photosynthesis can then begin and the plant can make its own organic food instead of relying on the food store of the seed. *Figure 1.10* and the accompanying text on page 8 are relevant to events at this stage. Seed dormancy, the role of plant growth substances during germination, and the physiology of germination are discussed in chapter 5.

Questions

1 Discuss how the structure of flowers is linked to their function.
2 Discuss the relative merits of self- and cross-pollination.
3 a What is a seed?
 b Describe the structure of a seed just before germination and the functions of its various parts.
 c Briefly describe the origins of each of these parts.

SUMMARY

■ Sexual reproduction is the fusion of two gametes to form a diploid zygote. It involves the fusion of the nuclei of the gametes.

■ The organs of sexual reproduction are produced in flowers in the flowering plants. Male gametes are produced in pollen grains inside anthers, and female gametes in the embryo sac which is inside the ovule. Meiosis occurs during the formation of pollen grains and embryo sacs.

■ Pollen must be transferred to the female parts of the flower, a process called pollination. Self-pollination or cross-pollination may occur. Both have their own particular advantages, and many plants have evolved special mechanisms to promote one or the other.

■ Cross-pollination produces more genetic variation because two parents are involved. Wind- and insect-pollination particularly favour cross-pollination.

■ Self-pollination is more reliable and less wasteful of pollen but, as only one parent is involved, inbreeding occurs.

■ To achieve fertilisation a pollen tube carrying the two male gametes must grow to the ovule.

■ In flowering plants a double fertilisation takes place. One male gamete fuses with the female gamete to produce the zygote. This develops into the embryo. The other male gamete fuses with a diploid nucleus to generate a triploid tissue, the endosperm.

■ In endospermous seeds, the endosperm becomes the food store. In non-endospermous seeds, the endosperm does not develop and the cotyledon or cotyledons (leaves of the embryo) swell with food to become the food store.

■ The fertilised ovule is the seed. There may be one or more seeds. They are enclosed in the ovary, which is known as the fruit after fertilisation.

■ At germination, the cotyledons may remain below ground (hypogeal germination) or be carried above ground (epigeal germination).

Sexual reproduction in humans

For humans, like other animals, the drive to reproduce is one of the most basic and important that we have. We are typical mammals; the fetus develops inside the mother, nourished by the placenta, and has an extended period of development. In this chapter and in part of chapter 5 we will consider some of the biological facts of reproduction, particularly the important role of hormones. But we will also explore some of the important social issues associated with reproduction. Our ability to 'interfere' with or intervene in the process is unique to our species and is increasing all the time. Many important moral, ethical, legal and social issues are raised as a result. Although biologists and the medical profession cannot be expected to have all the answers to the problems raised, they are essential contributors to the inevitable debate within society.

If the female and male reproductive systems are examined by dissection, it is found that the urinary (excretory) system is very closely linked with the reproductive system. This is especially true in the male, so traditionally both systems are studied together as the **urinogenital system (UG system)**. The diagrams used here are therefore of the UG systems.

The female reproductive system

Figures 4.1 and *4.2* give a realistic impression of the structure of the female UG system. *Figure 4.3* is a simplified diagram. There are two almond-shaped **ovaries**, about 2.5–5 cm long and 1.5–3 cm wide. They are the site of production of the female gametes, called **ova** (eggs), and they also produce the female sex hormones **oestrogen** and **progesterone**. The **oviducts** (also called the **Fallopian tubes**) end in funnels fringed with feathery processes called **fimbriae**. They collect secondary oocytes released by the ovaries. Their lining is ciliated and muscular, and the movements of the cilia and muscles sweep the egg towards the uterus. Fertilisation takes place in the oviduct. The **uterus** is about 7.5 cm long and 5 cm wide (about the size and shape of an inverted pear), but it can grow three to six times larger during pregnancy. It lies behind the bladder and has a thick muscular wall (the **myometrium**) and a well-blooded lining (the **endometrium**) which is shed during menstruation. The **cervix** (the neck) is the narrow junction between the uterus and the vagina. The ring of muscle in the cervix can close off the uterus.

The **vagina** is a muscular tube about 8–10 cm long. It is the site where semen is deposited from the penis during sexual intercourse and also the birth canal during childbirth. The walls contain elastic tissue and the lining (epithelium) is folded.

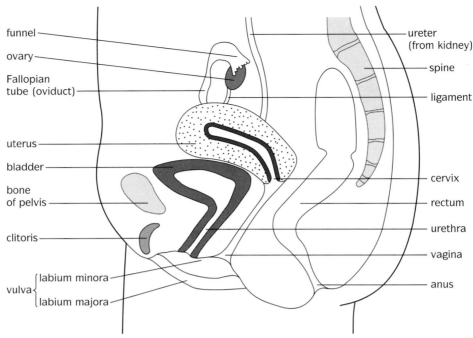

● **Figure 4.1** Side view of female urinogenital system (in section).

It can therefore enlarge to allow the entry of an erect penis or the exit of a baby. The epithelium secretes acids which deter the growth of harmful microorganisms. Acids are harmful to sperm, hence the need for semen to be alkaline.

The external genital organs (the **vulva**) include the labia majora, the labia minora, secretory glands and the clitoris. The **labia majora** are two longitu-

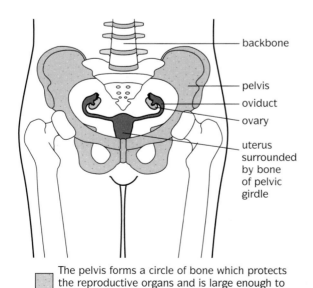

The pelvis forms a circle of bone which protects the reproductive organs and is large enough to accommodate the uterus during pregnancy.

● **Figure 4.2** The pelvic cavity in the female

dinal folds of skin containing fat, smooth muscle and many sensory receptors. The **labia minora** are two smaller folds of skin lying between the labia majora, also with many sensory receptors. The labia protect the openings of the vagina and urethra. The **clitoris** is situated at the top of the vulva, just below the point where the labia minora meet. It is small (less than 2.5 cm long), the female equivalent of the penis. Like the penis, it contains many nerve endings and swells with blood, becoming erect when sexually stimulated. It is a major source of sexual arousal during sexual intercourse.

The male reproductive system

Figure 4.4 gives a realistic impression of the structure of the male UG system from the side. *Figure 4.5* is a simplified diagram drawn from the front. Note that, unlike the female, the male has a shared external opening of the urinary and reproductive systems. Both the bladder and the vasa deferentia open into the urethra, which therefore carries both urine and semen.

The **testis** (plural **testes**; there are normally two present) is the site of production of the male

Urinary system

Reproductive system

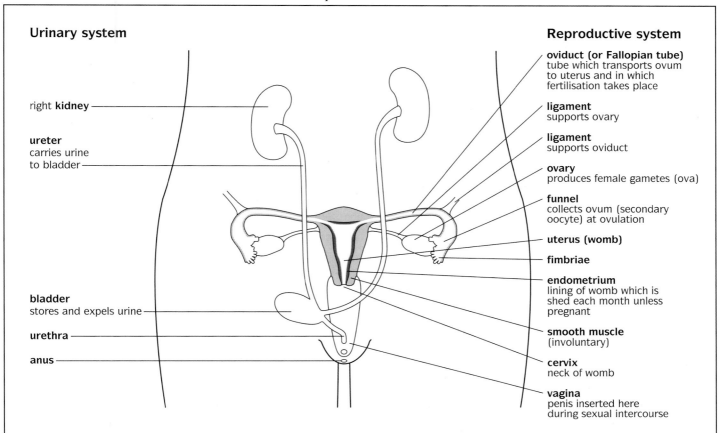

right **kidney**

ureter
carries urine
to bladder

bladder
stores and expels urine

urethra

anus

oviduct (or Fallopian tube)
tube which transports ovum
to uterus and in which
fertilisation takes place

ligament
supports ovary

ligament
supports oviduct

ovary
produces female gametes (ova)

funnel
collects ovum (secondary
oocyte) at ovulation

uterus (womb)

fimbriae

endometrium
lining of womb which is
shed each month unless
pregnant

smooth muscle
(involuntary)

cervix
neck of womb

vagina
penis inserted here
during sexual intercourse

● *Figure 4.3* Female urinogenital system. The bladder is shown moved to one side to reveal the
uterus. Note that the urinary and reproductive systems have completely separate openings.

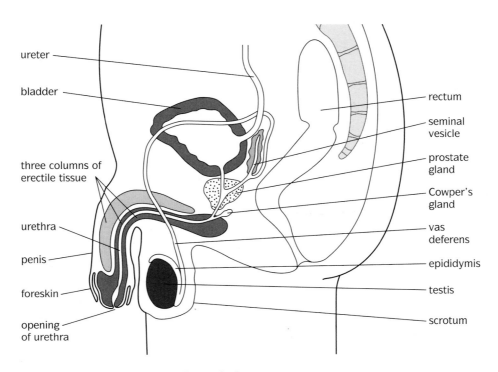

ureter

bladder

three columns of
erectile tissue

urethra

penis

foreskin

opening
of urethra

rectum

seminal
vesicle

prostate
gland

Cowper's
gland

vas
deferens

epididymis

testis

scrotum

● *Figure 4.4* Side view of male urinogenital system (in section).

gametes, the **spermatozoa (sperm)**. The **seminiferous tubules** are tightly coiled tubes (about 1000 per testis). Their combined length is about 225 metres. Cells lining their walls produce the sperm. The **epididymis** (again, there are two) is a coiled tube about 6 m long in which the sperm are stored while completing their maturation. Some fluid is reabsorbed, making the sperm more concentrated (about 5000 million/cm^3). It is also where the sperm become mobile. The **scrotal sac,** or **scrotum,** is a sac of skin containing the two testes which hangs from the main body cavity and helps to keep the sperm about 3 °C cooler than normal body temperature. This improves their survival rate. The **vas deferens** (plural **vasa deferentia**; there are two of them) is a tube which carries the sperm out of the testis to the **urethra.** Sperm are also stored there.

The **prostate gland, Cowper's glands** (paired) and **seminal vesicles** (paired) are all glands which secrete fluid for carrying the sperm and in which the sperm can swim. Fluid plus sperm is called **semen.** The fluid of the seminal vesicles (**seminal fluid**) and Cowper's glands is alkaline and neutralises the acidity of any remaining urine in the urethra. The prostate gland secretes mucus, and the seminal

vesicles secrete fructose, a sugar used by sperm for energy. Other chemicals, probably involved in activation of sperm, are also secreted by the glands.

The **penis** contains the urethra, which carries sperm to the outside world. The penis also contains special spongy tissue which can fill with blood when the male is sexually stimulated, causing it to become erect and rigid. The penis ejaculates semen into the vagina of the female during sexual intercourse.

Gametogenesis

We shall now examine the three main stages of reproduction, namely gametogenesis, fertilisation and development of the zygote. **Gametogenesis** is the formation of gametes. It takes place in the **gonads,** which are the testes in the male and the ovaries in the female. It involves meiosis in the nuclei of diploid 'mother cells' to form haploid gametes. The importance of halving the chromosome number in this way was stressed in *figure 2.7* of *Foundation Biology* in this series. It means that when a male gamete fuses with a female gamete, the normal diploid number is restored. It also has the important consequence of increasing genetic variation.

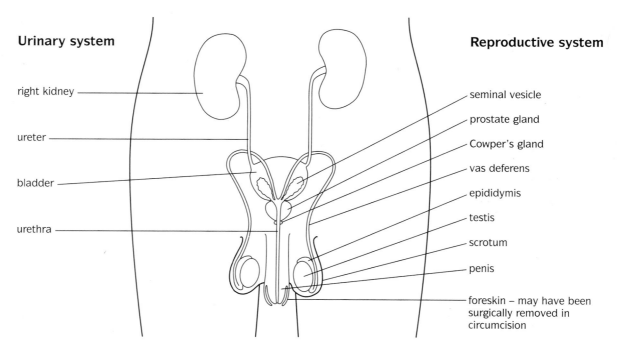

Urinary system

right kidney

ureter

bladder

urethra

Reproductive system

seminal vesicle

prostate gland

Cowper's gland

vas deferens

epididymis

testis

scrotum

penis

foreskin – may have been surgically removed in circumcision

● *Figure 4.5* Male urinogenital system.

Formation of sperm is called **spermatogenesis**. Formation of eggs is called **oogenesis**. (Technically, the female gametes are called **ova** (singular **ovum**) but biologists and the medical profession commonly refer to them as eggs.) The two processes are similar in essence. In each case, cells of the germinal epithelium, which is the outer layer of the ovary or seminiferous tubule, multiply before maturing into egg or sperm mother cells. The mother cells are known as **oocytes** and **spermatocytes**, respectively. Their nuclei then divide by meiosis to produce haploid gametes. The process is shown diagramatically in *figure 4.6*.

SAQ 4.1

Using the information provided in *figure 4.6*, **a** how many sperm can be formed from **(i)** one primary spermatocyte **(ii)** one secondary spermatocyte? **b** What is the name of an immature sperm?

Spermatogenesis

Unlike the female, the sexually mature human male produces vast numbers of gametes in a continuous production line of several thousand per second (over 100 million per day). This process of spermatogenesis takes place in the testes, shown in section in *figures 4.7* and *4.8*. Each testis contains a mass of tubes called the **seminiferous tubules** and it is in their walls that spermatogenesis takes place (summarised in *figure 4.6*). Sperm development takes place from the outer layer of the tube, the germinal epithelium, towards the centre of the tube where mature sperm break away from the wall and float down the tube towards the epididymis for storage. A ring of special cells, the **Sertoli cells**, is present in the wall. They secrete the fluid found in the lumen of the tubes. All stages of sperm development take place in close association with the Sertoli cells *(figure 4.7b)* and the complex modelling of the sperm, particularly spermatids, is done with their assistance. It involves phagocytosis of some of the spermatid cytoplasm.

Between the tubes are connective tissue, blood vessels and special **interstitial cells**, also known as **cells of Leydig** *(figures 4.7a* and *4.8)*, which secrete the male sex hormones, including **testosterone**. These hormones circulate in the blood and are important in the development of the male secondary sexual characteristics (those that develop at puberty). Testosterone also enters the seminiferous tubules and stimulates the cells, especially the Sertoli cells, which are involved in spermatogenesis.

Structure of the sperm

The structure of the sperm is beautifully linked with its function, as shown in *figure 4.9a*. The **acrosome** is a large, modified lysosome containing the hydrolytic enzymes

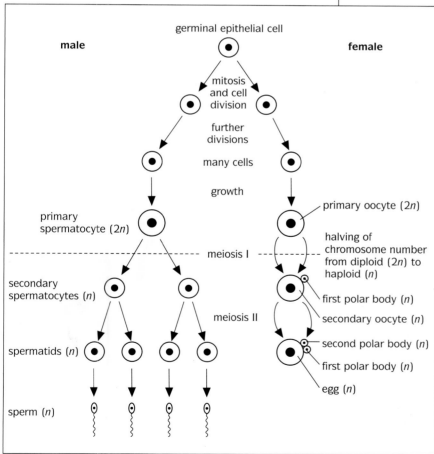

● **Figure 4.6** Diagrammatic representation of gametogenesis in the male and female.

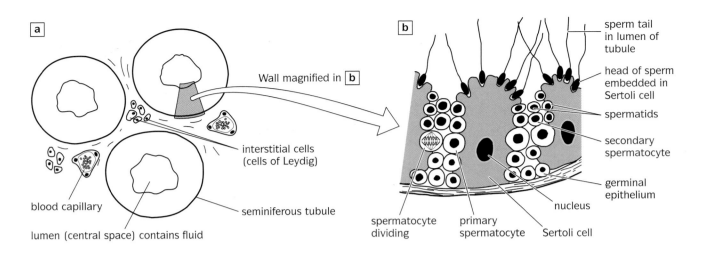

● *Figure 4.7* Diagram showing the microscopic structure of the testis.
a Group of three seminiferous tubules.
b Enlarged portion of the wall of a seminiferous tubule showing stages in sperm development.

needed to digest a path to the egg at fertilisation. The **nucleus** carries a haploid set of chromosomes, the genetic information from the male. The **axial filament** is made of microtubules and is responsible for the wave-like beating of the tail which propels it through its fluid surroundings at an average rate of 30 cm per hour. The microtubules have a characteristic '9 + 2' arrangement consisting of a ring of nine pairs of microtubules surrounding two central microtubules *(figure 4.9b)*. They arise from the **centriole**. The **middle piece** and **tail** are

concerned with propulsion. Numerous mitochondria provide the energy for beating the tail. This energy is contained in ATP made during aerobic respiration.

Oogenesis

Figure 4.10 shows the structure of a rabbit ovary as seen in section with a light microscope. Unlike the situation in the human, several Graafian follicles may develop at the same time in the rabbit. A diagram of a section through a human ovary is shown in *figure 4.11*. The process of egg formation, or oogenesis, starts before birth. The outer layer of the ovary, the germinal epithelium, produces **primary oocytes** *(figure 4.6)*. It also produces cells known as **follicle cells** which multiply and cluster around the oocytes, forming structures known as **primary follicles**. In humans about two million of these are already present at birth. By birth, each primary oocyte has started to divide by

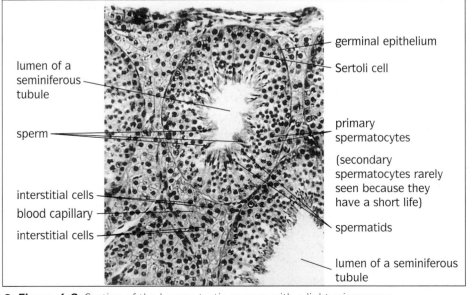

● *Figure 4.8* Section of the human testis as seen with a light microscope.

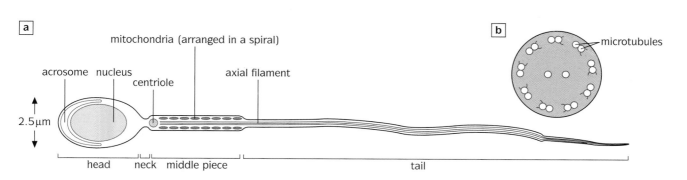

● *Figure 4.9*

a Structure of a human sperm. Total length is 60 μm.

b TS tail showing the '9 + 2' arrangement of microtubules.

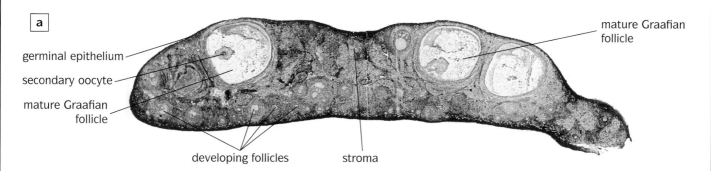

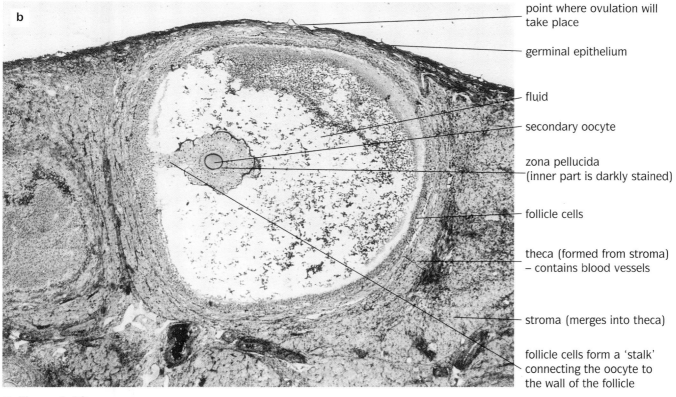

● *Figure 4.10*

a LS ovary of a rabbit showing developing follicles.

b High power detail of one mature Graafian follicle.

meiosis, but the process stops once the chromosomes pair up in prophase I. Once a human female is sexually mature, usually one primary follicle per month is stimulated to complete development into a **Graafian follicle**. This is surrounded by a layer called the **theca** which secretes the hormone **oestrogen**. The follicle cells multiply and several fluid-filled spaces develop which eventually fuse to form one space (*figures 4.10* and *4.11*). The primary oocyte completes meiosis I and divides unequally into two. The smaller daughter cell is known as a **polar body** and eventually disintegrates. The larger daughter cell is known as the **secondary oocyte** and this, together with some of the surrounding follicle cells, is released at ovulation. The Graafian follicle is large enough (about 1 cm in diameter) to cause a blister-like swelling on the surface of the ovary at ovulation (*figure 4.11*). This can be seen using a laparoscope (see page 67). After ovulation the empty follicle develops into a **corpus luteum** (yellow body), which gradually degenerates unless fertilisation takes place. The corpus luteum secretes the hormone **progesterone** (see below).

Structure of the secondary oocyte and egg

The structure of the secondary oocyte and egg is shown in *figure 4.12*. The egg is the largest cell in the human body (140 μm or 0.14 mm in diameter) and is just visible with the naked eye. Unlike the egg of many animals it contains no conspicuous yolk, although it does have a lipid food reserve. It also obtains nutrients from surrounding cells. The lysosomes function at fertilisation.

SAQ 4.2

List the main differences between a human sperm and egg, giving brief reasons for the differences. What important similarity do they have?

Hormonal control of gametogenesis

There are certain important similarities between the hormonal control of spermatogenesis and oogenesis. In both cases the control centres are the

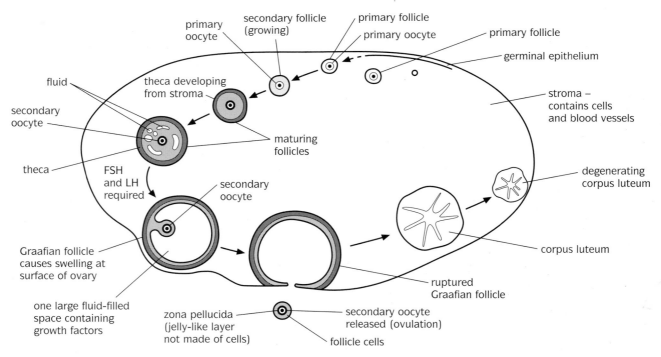

● *Figure 4.11* Stages in the development of one follicle in a human ovary. Arrows show sequence of events.

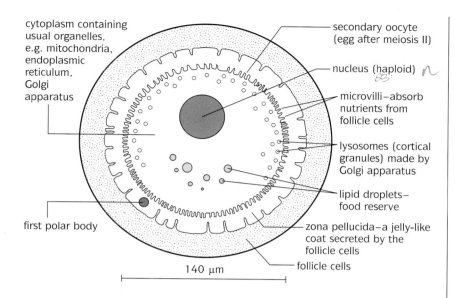

cytoplasm containing usual organelles, e.g. mitochondria, endoplasmic reticulum, Golgi apparatus

secondary oocyte (egg after meiosis II)

nucleus (haploid)

microvilli–absorb nutrients from follicle cells

lysosomes (cortical granules) made by Golgi apparatus

lipid droplets– food reserve

first polar body

zona pellucida–a jelly-like coat secreted by the follicle cells

follicle cells

140 μm

● **Figure 4.12** Structure of the secondary oocyte and surrounding structures at ovulation. At fertilisation, the secondary oocyte divides (meiosis II) to form the egg and a second polar body. The egg has the same structure as the secondary oocyte.

hypothalamus and the **pituitary gland**. The pituitary gland secretes two **gonadotrophic hormones, follicle stimulating hormone (FSH)** and **luteinising hormone (LH)**. The term gonadotrophic means that they stimulate gonads. The gonadotrophic hormones themselves are secreted in response to a hormone signal from the hypothalamus. This hormone is **gonadotrophin releasing hormone (GnRH)** and is transported in a special blood vessel linking the hypothalamus with the anterior lobe of the pituitary gland. GnRH must be present for normal gonad function, and is also the means through which the central nervous system can influence reproduction. The hypothalamus and pituitary gland act as the controlling link between the nervous system and the endocrine system.

Regulation of gonadotrophins also involves negative feedback from hormones produced by the gonads. In order to understand the following detailed accounts, it will be useful to recall the principles of negative feedback by referring to page 71 of *Foundation Biology*.

Hormonal control of spermatogenesis

The hormonal control of spermatogenesis is summarised in *figure 4.13*. Spermatogenesis is dependent on the production of testosterone by the testes. In turn, production of testosterone is controlled by the pituitary gland. This can be deduced from the observation that if the pituitary gland is removed from a rat, its testes shrink in size, spermatogenesis ceases and testosterone levels

decline. If doses of testosterone are then given to the rat, the changes can be partly reversed. The anterior lobe of the pituitary gland achieves this control by secreting LH. This hormone is also secreted by the female. In the male it has an alternative name, **interstitial cell stimulating hormone (ICSH)**. It travels in the blood from the pituitary to its targets, which are specific receptors on the membranes of the interstitial cells of the testes. Here it stimulates the production of testosterone from the steroid, cholesterol. Testosterone then stimulates spermatogenesis in the seminiferous tubules. Further experiments have revealed that, for completely normal levels of sperm production, another pituitary hormone, FSH, is also needed. FSH binds to specific receptors on the membranes of Sertoli cells and makes them much more active, stimulating sperm development and the secretion of fluid into the seminiferous tubules. Regulation of LH is brought about by testosterone in a process of negative feedback. Regulation of FSH is thought to be achieved by a product of the Sertoli cells known as **inhibin**.

The menstrual cycle and hormonal control of oogenesis

Figure 4.14 summarises the hormonal control of the menstrual cycle. In the male, the gametes and sex hormones are produced continuously, but in the female events are cyclic. There are two

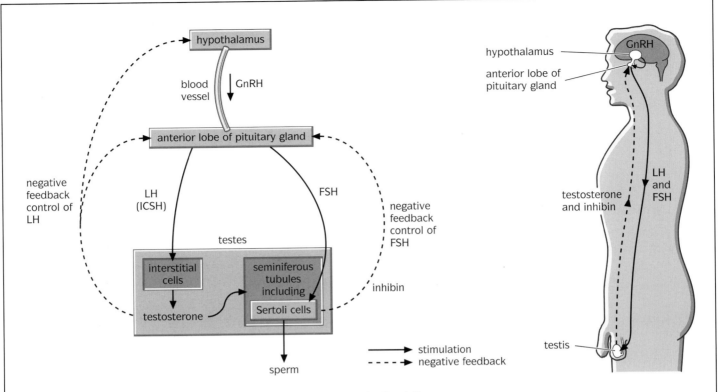

● **Figure 4.13** Summary of hormonal control of spermatogenesis. Seminiferous tubules, including Sertoli cells, must be active for successful production of sperm. In the male, hormone production is continuous, not cyclic as in the female.

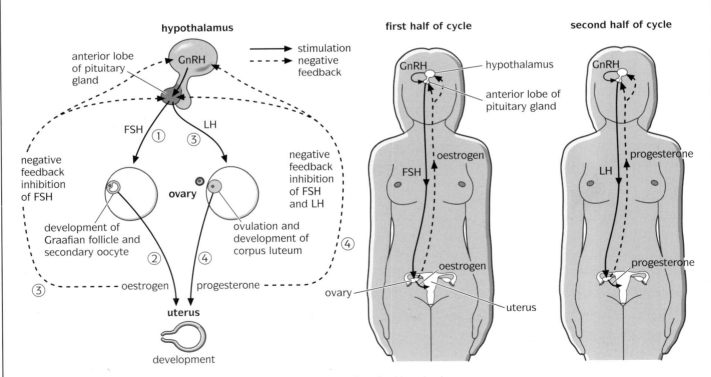

● **Figure 4.14** Summary of hormonal control of the menstrual cycle. Note both oestrogen and progesterone have a negative feedback effect on the hypothalamus as well as the anterior pituitary gland.

halves to the cycle. In the first half, an egg is produced; in the second half, the uterus is prepared for implantation in case the egg is fertilised. The complete cycle involves the ovaries, the uterus and the pituitary gland, and in humans is known as the **menstrual cycle**. The first day of the cycle is counted from the start of **menstruation**, which is the shedding of the bloody lining of the uterus (the endometrium) through the vagina. (Menstrual means monthly.)

The sequence of events is summarised as follows.

1 The anterior lobe of the pituitary gland secretes FSH into the blood. Its target is the **ovary**, where it stimulates development of a primary follicle.

2 The maturing follicle secretes oestrogen into the blood, increasing in amount as the follicle grows. Oestrogen has two main targets, the **uterus** and the **anterior lobe of the pituitary gland**. In the uterus it stimulates repair and, later, thickening of the endometrium. This involves the development of more blood vessels and glands. In the anterior lobe of the pituitary, oestrogen inhibits secretion of FSH and LH, an example of negative feedback. *Figure 4.15* shows how the production of oestrogen increases as the follicle grows, and

the relatively high levels just before ovulation. When levels of oestrogen increase greatly over a period of at least two days, LH and FSH secretion is *stimulated* rather than inhibited. Thus oestrogen has a double role in the pituitary. The whole process is very finely tuned.

3 LH and FSH are produced in a 'surge', a relatively large amount in a relatively short time. The target of LH is the ovary. It causes **ovulation**, which is defined as the release of the secondary oocyte from the Graafian follicle. The surge ensures the precise timing of ovulation, which usually occurs within 24 hours of the surge and, on average, 14 days into the cycle *(figure 4.15)*. LH also stimulates the remains of the Graafian follicle to develop into the corpus luteum, stimulates secretion of another hormone, progesterone, and reduces oestrogen production, which in turn reduces FSH and LH.

4 The corpus luteum secretes progesterone which has two targets, the **uterus** and the **anterior lobe of the pituitary gland**. In the uterus it stimulates glandular activity in the endometrium, and helps oestrogen in maintaining the thickness of the uterus. In the pituitary, high levels of progesterone enhance the negative feedback effects of oestrogen and inhibit FSH and LH secretion. The absence of FSH prevents another egg being fertilised while the female is pregnant, since a second egg cannot develop. Release of progesterone is associated with a temperature rise in the female body *(figure 4.16)*. Accurate measurement of body temperature at the same time each day can be a useful and reliable sign that ovulation has recently occurred.

5 If the woman becomes pregnant, the corpus luteum receives a hormone stimulus from the implanted embryo, but if the stimulus is *not*

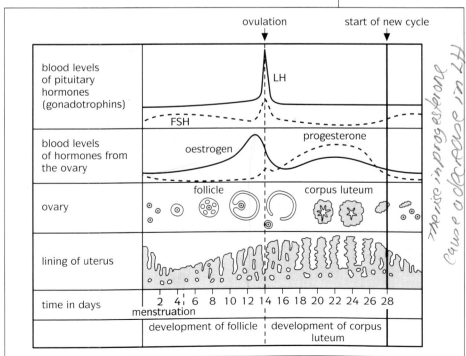

● *Figure 4.15* Summary of changes during the menstrual cycle.

The rise in progesterone cause a decrease in LH

The lowering of the amount of LH will increase the amount of progesterone

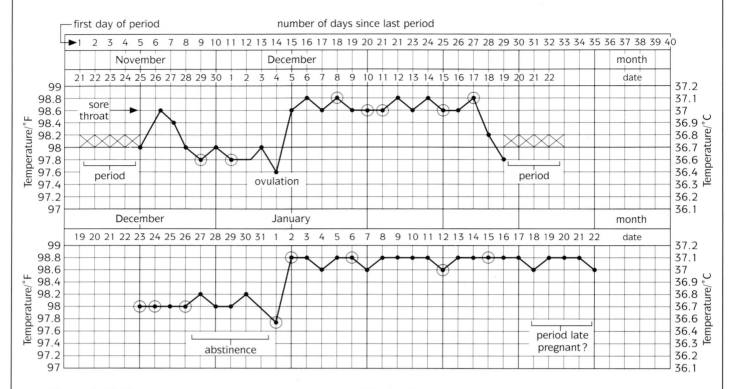

● **Figure 4.16** Temperature chart for a woman who was receiving fertility advice and trying to get pregnant. Occasions when sexual intercourse took place are circled. Temperature is now measured in degrees Centigrade in this situation.

received, the corpus luteum starts to degenerate. The reason for this is uncertain, but is probably due to a chemical produced by the corpus luteum itself. Pharmaceutical companies are interested in this possibility because such a chemical might be useful as a medical drug to induce early abortions. As the corpus luteum degenerates, the supply of oestrogen and progesterone is cut off. FSH is no longer inhibited and is therefore switched on, thus completing the cycle. The lining of the uterus breaks down at this stage, causing menstruation. This typically continues for about the first five days of the next cycle (*figure 4.15*).

SAQ 4.3

Study *figure 4.16*. **a** Why is it important to take the temperature at the same time every day? **b** Why would the woman be advised to record any illness, such as a sore throat, on the chart? **c** What is the normal temperature **(i)** before ovulation? **(ii)** after ovulation? **d** What effect does pregnancy have on temperature? **e** On what date(s) is she most likely to have conceived?

Passage of sperm from testes to oviduct

Transport in the male

For fertilisation to take place, sperm must travel from the seminiferous tubules, where they are made, to the oviduct in the female. The seminiferous tubules are grouped into bundles of about 100. From each bundle one tube emerges which connects the bundle to the epididymis. Each sperm spends about six to twelve days moving with the fluid from the seminiferous tubule and through the epididymis. Chemical changes in the fluid activate the sperm and they become motile. They are moved by muscular activity of the walls of the tubes from the epididymis into the vas deferens. At this stage they would be capable of fertilising an egg, but more fluid is added from the seminal vesicles, prostate and Cowper's glands (see page 42) to form **semen**, the mixture of sperm and seminal fluid. The combined fluids probably increase fertility.

Sexual intercourse

The sperm are introduced into the female during sexual intercourse (also called coitus or copulation). The fascinating related topics of social interaction, courtship and sexual behaviour are outside the scope of this book. Much of the biology of sexual behaviour is still poorly understood. For example, the effects of stress, social interaction and other environmental factors on fertility are important but, until recently, relatively neglected areas of research.

Sexual excitement of the male, either psychological or physical, results in erection of the penis. Erection is caused by dilation of arteries entering the penis and the flow of arterial blood into a special spongy tissue. In addition, the veins leading out of the penis constrict, resulting in raised blood pressure and higher blood volume in the penis. Sexual stimulation eventually comes to a climax with orgasm, an explosive wave of intense pleasure lasting a few seconds, accompanied by contraction of the muscles of the prostate gland, seminal vesicles, vas deferens and urethra. This causes ejaculation of the semen. A total volume of about $3\,cm^3$ of semen can be introduced into the vagina of the female. Research shows that coitus lasts an average of four minutes in the western world. It also suggests that many women do not achieve orgasm in this time. During coitus in the female, the clitoris may become erect and the blood supply to the vagina increases. Fluid from the blood seeps through the vaginal epithelium and lubricates the action of the penis. The vagina also expands. At orgasm, muscular contractions of both vagina and uterus occur. Orgasm can be centred on the clitoris or the vagina, the latter usually requiring longer to achieve.

Transport in the female

The sperm are deposited at the top of the vagina near the cervix. They can survive for one to two days in the female. The alkaline semen helps to protect them from the acid fluid (pH 5.7) of the vagina. Most of the sperm are thought to leak from the vagina without penetrating the cervix. The cervix is blocked by mucus and this is only thin enough to allow passage of sperm during the first part of the menstrual cycle, before progesterone levels become high. It is thought that movement of the sperm through the uterus and into the oviducts does not depend on muscular contractions of the uterus, but partly on their own swimming and partly on the action of cilia lining the uterus and oviducts. It probably takes at least four to eight hours for living sperm to reach the oviducts.

Capacitation

If freshly ejaculated sperm are mixed with eggs in the laboratory, as they are with in vitro fertilisation (the 'test tube baby' technique, see page 67), fertilisation is not possible for a few hours. The process which they must first undergo which gives them the capacity to fertilise an egg is called **capacitation**. It takes about seven hours. It involves the removal of a layer of glycoprotein and plasma proteins from the outer surface of the sperm. Glycoprotein is added by the epididymis, and plasma proteins come from the seminal fluid. They are normally removed by enzymes in the uterus (*figure 4.17*).

● *Figure 4.17* Scanning electron micrograph of human sperm in the uterus. The ciliated epithelium of the endometrium is visible. Capacitation occurs here.

Acrosome reaction

Once capacitation has taken place, sperm can be activated to finally achieve fertilisation. This takes place in the oviduct and is triggered by chemicals secreted by follicle cells or the zona pellucida around the egg. It includes a stage known as the **acrosome reaction** in which the acrosome in the sperm head swells and its membrane fuses in several places with the cell surface membrane surrounding the sperm head. This allows the enzymes inside the acrosome to escape *(figure 4.18)*.

Fertilisation

After ovulation, the secondary oocyte and surrounding cells are swept into the funnel at the end of the oviduct *(figure 4.3)* and along the oviduct by the action of cilia lining the funnel and oviduct. The secondary oocyte and its surrounding cells become surrounded by sperm *(figures 4.18 and 4.19)*. The process leading to fertilisation is described in *figure 4.18*. **Fertilisation** is the fusion of the sperm with the egg to form a single diploid cell, the zygote. It takes place in one of the oviducts. Penetration of the follicle cells surrounding the oocyte is shown in *figure 4.20*.

SAQ 4. 4

a Using *figure 4.18*, how is penetration of more than one sperm into the egg prevented?
b Why is it important that only one sperm enters the egg?
c Why are such large numbers of sperm produced if only one is needed for fertilisation?

Sperm may survive for up to three days in the woman's body, but they gradually lose their ability to fertilise the egg. After fertilisation, the zygote nucleus starts to divide by mitosis, thus starting development of the embryo. The first division occurs about 30 hours after insemination. The embryo reaches the uterus about four days after ovulation and becomes attached to the uterus wall, a process called **implantation**. This is complete by six

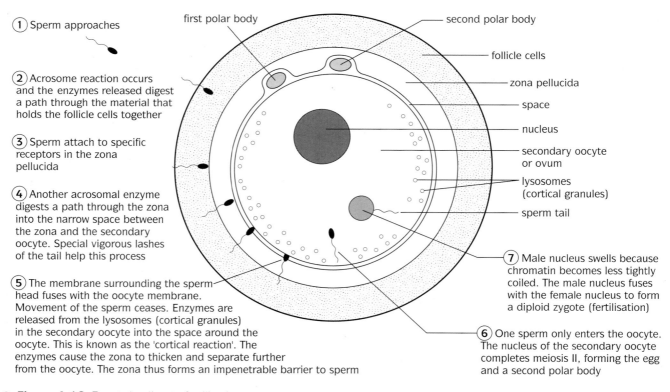

1 Sperm approaches

2 Acrosome reaction occurs and the enzymes released digest a path through the material that holds the follicle cells together

3 Sperm attach to specific receptors in the zona pellucida

4 Another acrosomal enzyme digests a path through the zona into the narrow space between the zona and the secondary oocyte. Special vigorous lashes of the tail help this process

5 The membrane surrounding the sperm head fuses with the oocyte membrane. Movement of the sperm ceases. Enzymes are released from the lysosomes (cortical granules) in the secondary oocyte into the space around the oocyte. This is known as the 'cortical reaction'. The enzymes cause the zona to thicken and separate further from the oocyte. The zona thus forms an impenetrable barrier to sperm

first polar body
second polar body
follicle cells
zona pellucida
space
nucleus
secondary oocyte or ovum
lysosomes (cortical granules)
sperm tail

7 Male nucleus swells because chromatin becomes less tightly coiled. The male nucleus fuses with the female nucleus to form a diploid zygote (fertilisation)

6 One sperm only enters the oocyte. The nucleus of the secondary oocyte completes meiosis II, forming the egg and a second polar body

● *Figure 4.18* Events leading to fertilisation.

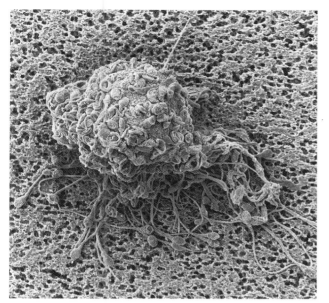

● *Figure 4.19* Scanning electron micrograph of sperm around an egg, just before fertilisation.

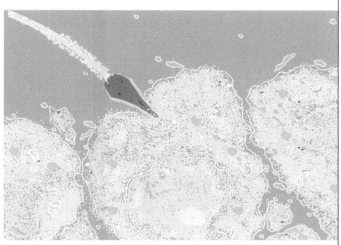

● *Figure 4.20* Transmission electron micrograph of an early stage of fertilisation showing a human sperm beginning to penetrate the layers of follicular cells that surround the oocyte. The nucleus (dark) is covered by the acrosome (light).

days after ovulation. A hormone produced by the embryo, **chorionic gonadotrophin (CG)**, prevents degeneration of the corpus luteum, and is the signal to the ovary that the woman is pregnant.

SAQ 4.5
Why is maintaining the corpus luteum important if pregnancy is to proceed?

Contraception

Humans, along with a few apes, are the only vertebrates that are sexually active throughout the reproductive cycle. This may be because sexual activity has evolved another role, that of pair bonding and reinforcing the emotional relationships between couples. However, for various reasons, it is usual for people, especially in modern societies, to want to limit the number of children they have. In the strict sense of the word, contraception means taking action to avoid conception, but, as will be seen, there are other procedures generally regarded as contraception which result in the early death of embryos. This raises ethical issues for some people. For cultural or religious reasons, some people are opposed to contraception, or unnatural methods of contraception. For instance, natural methods are the only forms of contraception officially allowed by the Roman Catholic Church. The main forms of contraception, their modes of action and their relative advantages and disadvantages are summarised in *table 4.1*. As the table reveals, there is at present no ideal contraceptive, so there is constant research into better methods. There has, for example, been much research into a male equivalent of 'the pill', a hormonal contraceptive pill or injection, which, like the pill used by women, is easy to take, virtually 100% reliable and has minimal side-effects. Starting in 1986, the World Health Organisation funded trials in which men received testosterone injections once a week to inhibit spermatogenesis. Results showed the method was very effective. A testosterone nasal spray and an implant (a device for slow release into the blood) are now being researched. It is anticipated that by the late 1990s an acceptable product will be available.

SAQ 4.6
Suggest how testosterone may inhibit spermatogenesis.

SAQ 4.7
Suggest reasons why a pharmaceutical company might be
a reluctant and **b** keen to develop a male pill.

Method	Mode of action	Effectiveness* if used correctly	Particular advantages	Particular disadvantages
Natural methods			no physical side-effects. Natural	
abstinence	avoid sexual intercourse	100%	totally effective and simple	restricts emotional development of a relationship
periodic abstinence (3 methods)	avoid intercourse around ovulation	85–93%	allowed by Roman Catholic Church	requires abstinence for part of cycle. Requires good knowledge of body and record-keeping
1 *rhythm method*	woman uses calendar to chart cycle		temperature and Billings method more accurate than rhythm. Can be quite effective if both methods used	relies on regular cycle
2 *temperature method*	temperature rises by about 0.6°C at ovulation			
3 *Billings method*	note changes in vaginal secretions around ovulation from thick, cloudy and sticky to thin, clear and stretchy	75–90%		
withdrawal (coitus interruptus)	man withdraws penis before ejaculation			low success rate because semen leaks from penis before ejaculation. Frustration if climax not reached. Considerable willpower required
Barrier methods	block sperm from reaching egg or inactivate sperm chemically			
condom (sheath)	thin rubber sheath placed over erect penis, preventing escape of sperm into vagina	97% (evidence shows less successful in first year of use)	simple, cheap, quite reliable, easily obtained. Increasingly popular because it protects against sexually transmitted disease such as AIDS	care must be taken to avoid damage to condom and that no spillage of semen occurs on withdrawal. Interruption of love-making to fit sheath
female condom (Femidom)	thin rubber or polyurethane tube, closed at one end. Three times wider than a male condom. Has two flexible rings, one at each end, to keep it in place. Forms a lining to the vagina. The closed end fits inside the vagina and the open end stays outside, flat against the vulva.	relatively new, so not much data available. Expected to be same as male condom.	as male condom. Can be inserted any time before intercourse, and removed any time later. No odour and does not grip around the penis.	partly visible outside body
cap (diaphragm or Dutch cap)	a soft rubber cap which covers the cervix and prevents entry of sperm. Inserted before intercourse. Caps are of various sizes and design. Used with a spermicide cream which kills sperm. Must be left in place at least six hours after intercourse	85–97%	can be inserted a few hours before intercourse	training required to fit cap. Need to check every six months that cap is right size and correctly fitted. Occasionally causes abdominal pain.
sponge	sponge impregnated with spermicide. Fits over cervix. Disposable. Fit up to 24 hours before intercourse. Leave in place at least six hours after intercourse.	75–85%	easier than cap because one size fits all and no fitting required	relatively high failure rate

Method	Mode of action	Effectiveness* if used correctly	Particular advantages	Particular disadvantages
Sterilisation				
Male – vasectomy	each vas deferens is tied and cut by a surgeon	100%	very reliable. Simple. No side-effects. Semen still produced, but without sperm	very difficult to reverse
Female – tubal ligation	oviducts tied and cut by a surgeon	100%	very reliable	very difficult to reverse. Operation more difficult and possibility of reversal even less than in male
Hormonal methods	prevent ovulation – ovaries do not produce eggs (FSH switched off)		reliable and convenient	sometimes medical side-effects
combined pill	contains oestrogen and progesterone. Pill taken orally daily for 21 or 28 days each month	99–100%	convenient. Does not interfere with love-making. Extremely reliable. Gives regular periods. Reduces period pains	minor side-effects possible. Not suitable for older women or smokers over 35, or those with risk of heart disease or strokes. Serious side-effects possible (though rare). Available only on prescription
mini-pill	contains progesterone only. Ovulation may occur, but cervical mucus thickened and prevents entry of sperm	98%	lower dose of hormone and therefore less risk to older women. Can be taken while breast-feeding without affecting quantity or quality of milk	monthly cycle may be irregular, with breakthrough bleeding. Minor side-effects possible, e.g. headaches. Must be taken same time each day (or no more than three hours late)
injectable contraceptives, e.g. Depo-Provera	progesterone injected into muscle for long-term, slow release. Stops ovulation. Only recommended if other methods unsuitable	over 99%	one injection protects for up to 12 weeks	periods usually irregular. May delay return to fertility when stopped
Preventing implantation		.		
intra-uterine device (IUD or coil)	small device, usually plastic with copper, inserted into uterus by doctor and left in place. Prevents implantation	96–99%	effective for up to 5 years Particularly suitable for older women. Does not interfere with lovemaking	periods may be heavier at first. IUD may come out. May lead to pelvic infection. Often causes cramps and pain
morning-after pill	pill(s) containing high doses of hormone taken within three days of intercourse	high	used in emergency when intercourse has taken place without contraception	long-term effects unknown and probably harmful. Therefore cannot be used on a regular basis

* **Effectiveness**: 75% effectiveness means 25% failure; that is 25 pregnancies per 100 women using the method per year. Unprotected sex results in 90% failure.

● *Table 4.1* Contraception – methods, modes of action and relative advantages and disadvantages

SAQ 4.8

An anti-progesterone contraceptive is being developed. From your knowledge of the effects of progesterone before and after conception, what ethical arguments might there be against it?

Ethical issues

As a species, we are increasingly able to intervene in our own reproduction, for instance by contraception, abortion and in vitro fertilisation. Many important ethical issues arise from these examples, some global and some local. Ethics is the science of morals, the study of right and wrong. It has been said that it is the job of scientists to *solve problems*. Should scientists then be involved with ethical issues where we sometimes have to accept that there are no 'right' solutions, and that what is right for some may be wrong for others? In discussing these issues it can be argued that it *is* important to be well-informed scientifically. Biological knowledge, combined with the willingness to listen with respect to the views and experience of others, is essential in the debate.

The following are some of the ethical issues associated with contraception.

■ The Roman Catholic Church argues that the use of contraception is interference with a natural process of procreation which should be left to the will of God; the enjoyment of sex is regarded as a gift that should not be separated from the purpose of creation. However, having many children may condemn families to lives of poverty, especially in developing countries. Women may resort to abortion, either legal or illegal, as an alternative to contraception (see also the ethical issues associated with abortion on page 65).

■ Should more advice about contraception be given to young people? If so, at what age and by whom should the advice be given?

■ Should contraceptives be made more easily available? For example, should schools or colleges contain contraceptive machines?

■ Should methods which prevent implantation be avoided since a potential human life is destroyed?

■ Some contraceptives may be associated with long-term health risks, for example the pill and RU486, an anti-progesterone drug which aborts the fetus (see page 64).

■ Should research into male contraception be given a higher priority, and should men be encouraged to take more responsibility?

The placenta

The placenta is a structure found only in mammals. It is a unique organ in that it is formed from the tissues of two genetically different individuals, the mother and the fetus. Its function is to allow the exchange of materials between mother and fetus.

Structure

Figure 4.21 shows the structure of the placenta. Early in development a layer called the **chorion** develops from the embryo and invades the uterus wall, forming many finger-like projections called **chorionic villi**. These project into blood-filled spaces formed from expansion of the mother's blood vessels. Inside the villi, tree-like networks of fine blood vessels develop from the fetus. Villi and capillaries together provide a large surface area for exchange with the mother's blood. The epithelial cells of the villi have microvilli, increasing the surface area for uptake even more. Deoxygenated blood from the fetus, carrying waste products, is pumped by the heart of the fetus through the two umbilical arteries in the umbilical cord *(figure 4.22)* into the villi where the fine branches of the arteries form the capillaries. Substances such as nutrients and oxygen cross by a variety of methods from the maternal blood spaces through the thin walls of the villi and capillaries and return to the fetus in the umbilical vein *(figure 4.21b)*. Diffusion, facilitated diffusion, active transport and pinocytosis have all been observed. The fetal and maternal blood vessels are not directly connected, so fetal blood does not mix with maternal blood.

SAQ 4.9

Suggest reasons why there is no direct contact between maternal and fetal blood in the placenta.

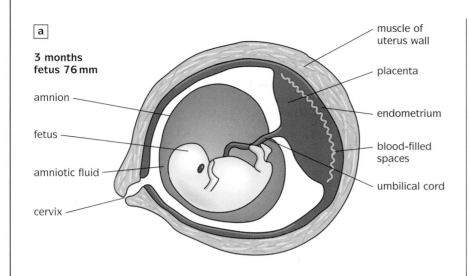

a

**3 months
fetus 76 mm**

amnion

fetus

amniotic fluid

cervix

muscle of
uterus wall

placenta

endometrium

blood-filled
spaces

umbilical cord

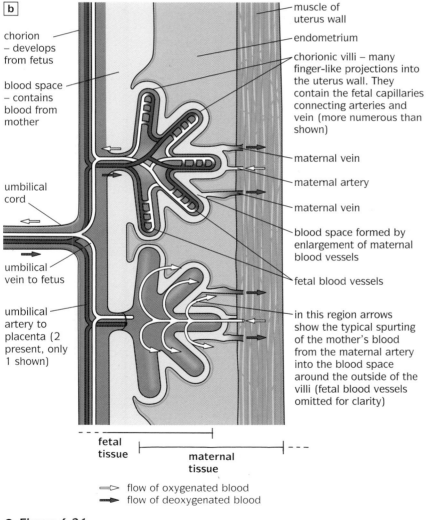

b

chorion
– develops
from fetus

blood space
– contains
blood from
mother

umbilical
cord

umbilical
vein to fetus

umbilical
artery to
placenta (2
present, only
1 shown)

muscle of
uterus wall

endometrium

chorionic villi – many
finger-like projections into
the uterus wall. They
contain the fetal capillaries
connecting arteries and
vein (more numerous than
shown)

maternal vein

maternal artery

maternal vein

blood space formed by
enlargement of maternal
blood vessels

fetal blood vessels

in this region arrows
show the typical spurting
of the mother's blood
from the maternal artery
into the blood space
around the outside of the
villi (fetal blood vessels
omitted for clarity)

fetal
tissue

maternal
tissue

⟹ flow of oxygenated blood
➡ flow of deoxygenated blood

● **Figure 4.21**

a Diagram showing how the umbilical cord and placenta link the fetus to the
mother's blood supply.

b Part of the placenta showing blood supply.

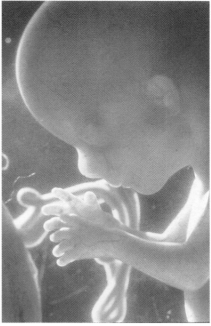

● **Figure 4.22** Human fetus at about
4 months (16 weeks) showing the
head and upper limbs and the
umbilical cord which connects the
fetus to the placenta. The eyelids
have not yet fully developed.

SAQ 4.10

Using *figure 4.21* to help you, list the
parts of the placenta which are
composed of **a** fetal tissue, and
b maternal tissue.

Overall, the placenta is a relatively
large, pancake-like structure. At
birth it is an average 15–20 cm in
diameter and 3 cm thick in the
centre. It weighs about 600 g,
about one-sixth the weight of the
fetus. It becomes detached from
the uterus wall at birth and is
delivered after the baby as the
'afterbirth'. By the end of pregnancy, about 10% of the mother's
blood passes through the placenta
for each circulation of the body.

Functions of the placenta

The main functions of the placenta are to exchange products between mother and fetus, to make hormones and to act as a protective barrier. The following products move across the placenta.

■ **Respiratory gases.** Fetal haemoglobin has a higher affinity for oxygen than adult haemoglobin since the molecule has two alpha and two gamma chains instead of two alpha and two beta chains. This is necessary because the fetus has to 'steal' oxygen from the mother's haemoglobin and also the concentration (partial pressure) of oxygen in the mother's blood is not as high as in the atmospheric air we breathe. Oxygen diffuses from a higher concentration in the maternal blood space to a lower concentration in the fetal blood, passing through the thin (only one cell thick) walls of the villi and the capillaries. Waste carbon dioxide diffuses in the opposite direction down its own diffusion gradient.

■ **Nutrients** and **water.** All essential nutrients pass from maternal to fetal blood, including glucose, amino acids, fatty acids, glycerol, some fats, salts and vitamins. Water can cross by osmosis, due partly to differences in blood pressure, and hence water potential, between mother and fetus. The placenta can store glucose as glycogen as an emergency reserve of carbohydrate. The movement of glucose is a good example of facilitated diffusion; it moves down a concentration gradient using a special carrier protein in the cell

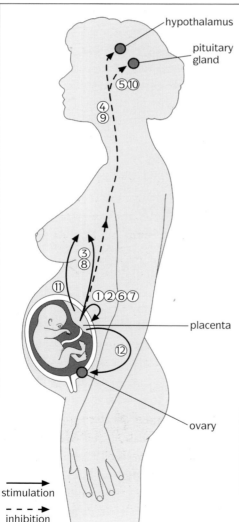

Oestrogen
① stimulates growth of uterus
② sensitises uterus to oxytocin
③ stimulates development of duct system of breasts
④ inhibits FSH
⑤ inhibits release of prolactin*, therefore inhibits lactation (secretion of milk)

Progesterone
⑥ maintains lining of uterus
⑦ relaxes muscle of uterus (prevents contraction and possible miscarriage)
⑧ stimulates development of milk glands in breasts ready for lactation
⑨ inhibits FSH
⑩ inhibits release of prolactin*

Human placental lactogen (HPL)
synthesis gradually increases during pregnancy
⑪ stimulates growth and development of breasts in preparation for lactation. Needed for oestrogen and progesterone to be effective (3 and 8)
Also adjusts glucose and fat metabolism of mother to advantage of fetus

Chorionic gonadotrophin (CG)
⑫ produced by the chorion from the time the embryo implants. Target is ovary. Maintains corpus luteum up to about 3 months, thus maintaining production of oestrogen and progesterone until placenta takes over this function. Level then declines.

* prolactin – some is synthesised before birth, but it is mainly released by the pituitary gland after birth; it stimulates milk secretion

● *Figure 4.23* Hormones secreted by the placenta and their functions.

surface membranes. Ions (such as sodium, potassium and calcium), amino acids, iron and vitamins probably cross by active transport. Sodium and potassium can also diffuse across the placenta.

■ **Excretory products.** Carbon dioxide is an excretory product of respiration. In addition, the fetus excretes nitrogenous waste, such as urea, which diffuses into the mother's blood and is removed by her kidneys.

■ **Antibodies.** The placenta allows the transfer of some antibodies from the mother to the fetus. Thus the fetus gains protection from the same diseases as the mother. This protection lasts for a few months after birth. However, problems can arise with the Rhesus blood group. About 84% of humans have Rhesus antigens in the cell surface membranes of their red blood cells (these are also found in Rhesus monkeys, hence the name). Such people are described as **Rhesus positive** (Rh+) whereas the 16% who lack the antigens are described as **Rhesus negative** (Rh−). If the mother is Rh− and the baby Rh+, there are circumstances, usually with the second or subsequent Rh− babies, when the mother will make Rhesus antibodies against the antigen. These can pass from mother to fetus across the placenta, attack the red blood cells of the fetus (leading to haemolytic disease of the newborn) and cause problems such as anaemia and jaundice.

SAQ 4.11

Give examples of substances which cross the placenta by the following mechanisms: **a** diffusion **b** active transport **c** facilitated diffusion **d** osmosis.

■ **Endocrine organ.** The placenta makes a number of hormones and is therefore part of the endocrine system. After three months of pregnancy it takes over from the corpus luteum as the main source of oestrogens and progesterone. *Figure 4.23* summarises the placental hormones and their functions.

■ **Protective barrier.** The placenta acts as a barrier to the passage of most bacteria. If the mother suffers a bacterial infection it is not therefore normally passed to the fetus. Some viruses, however, are small enough to cross the placenta. Two common problems are the rubella (German measles) virus and the HIV virus which can lead to AIDS. Mass vaccination has almost eradicated rubella in developed countries, but if the fetus *is* infected during the first eight weeks of pregnancy, the result can be blindness, deafness, heart problems or mental retardation.

Table 4.2 summarises some of the substances that cross the placenta.

The amnion

The **amnion**, like the chorion, is a structure which grows from the developing embryo (*figure 4.21a*). It is tough, thin and transparent and completely surrounds the embryo, and later the developing baby, throughout its development. The space enclosed, the **amniotic cavity**, becomes filled with a watery liquid called **amniotic fluid** in which the fetus is suspended. Its chief function is to protect the embryo or fetus from physical shock and damage. However, it also allows the fetus to move freely, which probably aids the development of muscles and bones. The amniotic fluid is similar in composition to the liquid part of the blood (serum)

	From fetus to mother	From mother to fetus
of benefit to fetus	CO_2, urea	O_2, nutrients, water, antibodies, antibiotics
potentially harmful to fetus	antigens, e.g. Rhesus antigen	some viruses, e.g. rubella, HIV; toxins; drugs, e.g. alcohol, crack (cocaine), heroin; tobacco products, e.g. nicotine, carbon monoxide; Rhesus antibodies

● *Table 4.2* Some of the substances that cross the placenta

of the fetus. There is constant exchange between the body of the fetus and the amniotic fluid via the respiratory and urinary pathways and the gut. Until it is about 20 weeks old, exchange can also take place through the skin of the fetus, which has not yet formed its protective dead layer. The fetus also swallows amniotic fluid (about $120 \, cm^3$ per hour at week 28). It urinates into the amniotic fluid (a peak of $26 \, cm^3$ per hour by week 40). The fluid may help to maintain a constant temperature around the fetus if the mother is subjected to extremes of temperature for long periods. The amniotic fluid is sampled during amniocentesis, a technique involving the removal of fluid by a long, hollow needle inserted through the abdominal wall. Chemical analysis of the fluid and examination of fetal cells, for example for chromosomal abnormalities, is a useful guide to certain disorders of the fetus.

Actions of the mother which affect fetal development

During the nine months of gestation the health and activities of the mother affect the fetus in many ways. This is particularly true in the critical period up to the end of the first three months when the major organ systems are being developed. At the beginning of this period the woman may not even know she is pregnant. After the first three months, the fetus mainly has to grow in size and mature. One of the aims of pre-natal (antenatal) care is to try to ensure that damage to the fetus is avoided. More recently, emphasis has also been placed on preconceptual advice because relevant life style changes take time to accomplish and should ideally be started before conception. Where possible, positive actions such as sensible levels of exercise, management of stress and healthy diet should be encouraged, as well as avoiding potentially harmful activities such as smoking. Some factors, such as housing quality, may be impossible to change. What, then, might a mother do that can affect fetal development? Some important factors will now be considered in more detail.

Nutrition

One old wives' tale suggests that the pregnant woman should 'eat for two'. If the size and growth rate of the baby relative to the mother are taken into account, 10% extra food intake might be a very rough guide, but *table 4.3* shows that, according to government guidelines of 1991, the situation is more complex than that.

SAQ 4.12

a What nutrients are needed in greater quantity in the diet during pregnancy? **b** In which cases are more than an extra 10% of normal requirements needed? **c** In which cases where more of a nutrient is needed is the extra **not** supplied by extra consumption? **d** In each of the cases in **a**, **b** and **c**, try to explain why the extra is required.

Smoking

Tobacco smoke contains three harmful ingredients, nicotine, tar (which contains carcinogens) and carbon monoxide. Nicotine and carbon monoxide can cross the placenta easily and are known to affect development of the fetus. (Remember that smoke enters the mother's lungs, not the lungs of the fetus, a common exam error!) If you speak to any midwife or nurse working in intensive baby care units, they will tell you how easy it is to recognise that a mother has been a smoker from the appearance of the placenta and umbilical cord at birth. The umbilical blood vessels are often narrower and the placenta smaller, helping to explain the fact that smoking mothers are more likely to miscarry and that their babies are often born underweight (on average 200 g underweight if the mother smokes 10–20 cigarettes per day). This phenomenon is common and known as **intra-uterine growth retardation (IUGR)**. Underweight babies are more at risk if there are birth complications, so smoking increases **perinatal mortality** (deaths just before, at, or just after birth). They are also more likely to be born prematurely and are less resistant to infection. Respiratory problems are more common (a sign of immaturity of the lungs). Smoking is also known to significantly reduce vitamin C uptake in the pregnant mother.

	Non-pregnant	Pregnant	Extra needed for / notes
energy/MJ	8.1	8.9*	growth of fetus; deposition of fat in mother's body for lactation later; but usually less physical activity late in pregnancy
protein/g	36	42	growth of fetus and maternal tissue, e.g. uterus, breasts
calcium/mg	525	525	bones and teeth and other purposes. During pregnancy effectiveness of calcium absorption increases, so no extra dietary calcium is normally needed.
iron/mg	11.4	11.4	extra haemoglobin for mother and fetus (pregnant woman has 1.5 l more blood). Anaemia and tiredness if deficient. Increased needs are achieved without extra in diet because none lost by menstruation, absorption from gut increases and stores are mobilised. Extra needed if low stores
vitamin A/μg	400	500	growth, development and differentiation of fetus, provision of reserves in liver of fetus, maternal tissue growth, particularly healthy skin and epithelia
thiamin (vitamin B$_1$)/mg	0.60	0.66	requirements related to energy needs – helps make energy available
riboflavin (vitamin B$_2$)/mg	0.9	1.2	release of energy from food
nicotinic acid (vitamin B$_3$)/mg	10.7	11.8	cell respiration – requirements related to energy needs
vitamin C/mg	25	35	healthy connective tissue. Stimulates absorption of iron
vitamin D**/μg	0	0	stimulates absorption and use of calcium, e.g. for bones and teeth
folate***/μg	150	250	efficient use of iron and therefore formation of haemoglobin/red blood cells. Pills commonly given as supplement during pregnancy. Lack leads to higher incidence of fetal neural tube defects such as spina bifida

* final 3 months only

** most people who go out in the sun need no dietary source of vitamin D since it is made
by the action of light on the skin

*** compounds derived from folic acid

● **Table 4.3** Estimated average requirements per day of nutrients for pregnant and non-pregnant women.

Carbon monoxide binds strongly to haemoglobin (mainly the mother's) to form carboxyhaemoglobin. This reduces the oxygen-carrying capacity of the mother, and hence the baby. This in turn slows the growth rate, resulting in an underweight baby. Recent evidence suggests oxygen transport across the placenta is accelerated by a carrier molecule and that carbon monoxide may reduce the efficiency of this carrier. **Nicotine** causes constriction of blood vessels and the most harmful consequence of this is probably the restriction of blood flow through the placenta, reducing the rate of exchange of nutrients, oxygen and other substances across the placenta. Nicotine also stimulates cardiac (heart) muscle and may make the baby's heart beat too fast.

Alcohol

There is a growing trend towards measuring alcohol intake in terms of grams (g). Measuring it in terms of *numbers* of drinks a day is meaningless since drink size is variable. To make it easier to calculate intake, 'standard units' are used. One unit equals 8 g of pure alcohol (10 cm^3). One glass of wine, or a short, can be considered as 1 unit, or 8 g of alcohol. This is roughly equivalent to half a pint of beer, although beers vary from 12 to 40 g of alcohol per pint.

Alcohol crosses the placenta easily. It is particularly damaging in the early stages of development, up to eight weeks. Quantities of 80 g or more per day (the equivalent of ten glasses of wine or five pints of beer or ten 'shorts') may give rise to a condition known as **fetal alcohol syndrome (FAS)**. This includes one or more of the following: mental retardation, reduced growth (before *and* after birth), poor muscle tone (the muscles are not in a state of partial contraction ready for action), heart defects, abnormal limb development, certain facial characteristics (such as a short, upturned nose, cleft palate, hare lip or receding chin) and behavioural problems. Growth of the infant continues to be slow after birth and problems such as hyperactivity or poor attention span and learning disorders continue in later years. Although this syndrome is regarded as extremely rare, it is not clear whether there is a *safe* level of alcohol. Some experts believe there is not. In 1988 a leading embryologist showed that alcohol may even damage eggs before conception, raising the chances of miscarriage and the incidence of chromosome abnormalities such as Down's syndrome. Effects can be far more subtle and harder to detect at lower consumption levels. How would you know if you had reduced the IQ of your child by 10 units? Other factors are often associated with excessive drinking, such as poor diet and poor living conditions, which make it difficult in practice to study the effects of alcohol in isolation. Late in pregnancy, alcohol may reduce circulation and exchange within the placenta, slowing down fetal development.

Moderate alcohol consumption seems to be associated with more miscarriages than normal, as well as low birthweight and congenital (at birth) malformations, especially of the nervous system (the brain in particular). For example, women who drink more than 100 g (about 12 units) of alcohol per week have more than twice the risk of delivering an underweight baby compared with those drinking less than 50 g (about 6 units) per week. Low levels of consumption (less than 50 g per week, but no more than 20 g (2.5 units) in a given day) seem to have no detectable consequences.

Other drugs

Most drugs can cross the placenta easily. Common pharmaceutical drugs such as aspirin, codeine, paracetamol and sleeping pills (barbiturates) cause no known damage to the fetus if recommended doses are taken. However, since drugs are likely to affect the fetus as well as the mother, doctors may advise avoiding drugs if possible. If the mother can become addicted to a drug, then so can the fetus. The classic and most dramatic example of a prescribed drug which did cause problems was thalidomide. This was introduced in the early 1960s as a remedy for morning sickness (nausea), which is particularly acute in some women. Within two years about 7000 children were born with severe physical handicaps as a result, limb development being particularly affected. The procedures for the testing of new drugs have since been changed to try to ensure that no similar tragedy occurs again. Women should always consult a doctor before taking medicines during pregnancy.

Drugs of abuse can have serious effects on the fetus. If the mother is dependent on a drug such as heroin or crack (cocaine), then her baby probably will be as well. Such babies are five times more likely to be born underweight and are twice as likely to be premature, with the associated dangers already discussed. Birth problems and miscarriages are more common. A drug-addicted baby is a pathetic sight and needs intensive care in a hospital. Although most survive, withdrawal symptoms occur over the first two weeks of life in 60–90% of babies born to drug-dependent mothers. Normal reflexes, such as the sucking reflex, are affected, the body may be rigid, uncontrolled tremors may be experienced and 'cot deaths' are more common.

Disease

As some disease-causing organisms can cross the placenta, particular care should be taken to avoid these. Of particular importance are the rubella, HIV and hepatitis B viruses. In the UK both hepatitis B and the HIV virus (which causes AIDS) are often associated with drug abusers and can be spread by contaminated syringes. Both can also be spread sexually. Vaccination against hepatitis B is expensive, and as yet there is no vaccination against the HIV virus. Sexually transmitted diseases can also be passed on to the baby as it passes through the vagina at birth.

Preconceptual and antenatal care

The mother-to-be can take positive steps to maintain or improve her own well-being and that of her unborn child by attending antenatal classes. Fathers are also encouraged to attend. The health of the baby and mother can then be monitored and relevant advice given. Antenatal care has led to a significant drop in perinatal mortality (that is, death at around the time of birth). More recently, preconceptual care for couples has been encouraged by the government and the medical profession, so that relevant advice on topics such as giving up smoking, health checks and genetic counselling can be given before pregnancy begins. It is hoped that the incidence of malformations of babies in England and Wales will thus be reduced from the current level of 2.5 per 100 live births.

Abortion

Abortion is the premature termination of pregnancy, resulting in death of the embryo or fetus. It may happen naturally, in which case it is generally called a **miscarriage** or **spontaneous abortion** or **natural abortion**. On the other hand, when carried out deliberately, it may be called **induced abortion**. In this book the term abortion is used to mean induced abortion.

The Abortion Act, which made abortion legal in the UK under certain circumstances, came into operation in April 1968. Since then more than three million abortions have been carried out in the UK, about four-fifths on UK residents and the rest on women from other countries. Overall about 20% of conceptions in the UK are terminated by abortion (in 1990, 8% of conceptions within marriage and 36% of conceptions outside marriage).

The Abortion Act was amended by the Human Fertilisation and Embryology Act (HFE Act), which came into force in April 1991. Two important changes were made. Firstly, the maximum age at which the fetus could be aborted was reduced from 28 to 24 weeks, with the exception described later. Twenty-four weeks represented the age at which the fetus could be expected to survive outside the mother's womb with medical assistance. It is the limit which receives most support from the public, although to a biologist it is a fairly arbitrary dividing line because the main problem preventing independent survival of the fetus just before 24 weeks is only the immaturity of its lungs. The second important change in the Act was to allow abortion at any stage of pregnancy in cases of severe fetal abnormality or serious risk to the health or life of the mother. Fifty-two abortions at over 24 weeks were carried out on these grounds in 1991 (between April, when the Act came into force, and December). The full legal grounds for abortion are as follows:

- risk to the woman's life;
- risk of grave permanent injury to the physical or mental health of the woman;
- risk of injury to the physical or mental health of the woman (up to 24 weeks) (most commonly used grounds);
- risk of injury to the physical or mental health of existing children (up to 24 weeks);
- substantial risk of a child being born seriously handicapped (anytime);
- emergencies (rare) (anytime).

The most important reason given for legalising abortion in 1968 was that it would reduce the number of illegal abortions. This, it was argued, would reduce the health risks to, and financial exploitation of, women and would also avoid the danger of the law being 'brought into disrepute' by being constantly ignored. These are partly ethical arguments, but other ethical issues are also raised which become matters for individual conscience rather than the law. These will be examined later.

SAQ 4.13

a From *table 4.4*, in what maternal age group do most abortions occur? **b** Plot bar charts of the data in *table 4.4*. **c** Calculate the percentage of women in each age category.

Age of women	Number in 1992
under 16	3 000
16–19	27 585
20–24	49 051
25–29	38 429
30–34	23 870
35–39	13 252
40–44	4 844
45+	452
not stated	12

● **Table 4.4** Age of women at time of abortion in 1992 (residents of England and Wales)

Five methods of abortion are in general use, the method used in a particular case depending mainly on the stage of pregnancy. The first four methods below involve a local or general anaesthetic, except the third method, where a general anaesthetic cannot be used.

■ **Vacuum aspiration** – up to 12 weeks of pregnancy. This is the most common method and is relatively safe and easy, taking about 30 minutes. The cervix is stretched, a narrow, flexible tube is inserted into the uterus and the fetal material is gently sucked out with a pump (aspirator). A variation of the technique, menstrual aspiration, takes only about five minutes and can be done up to six weeks after conception. A smaller tube can be used, so dilation of the cervix is not necessary. In this case the uterus lining is gently sucked out with a small syringe.

■ **Dilation and curettage (D + C)** – at 12–16 weeks of pregnancy. The cervix is dilated and the lining of the uterus scraped with a spoon-shaped cuvette (knife) to remove the fetus. The contents can be finally sucked out using vacuum aspiration (D + E) *(figure 4.24)*.

■ **Prostaglandins** or **saline injection** – at 16+ weeks of pregnancy. Prostaglandins or saline solution are injected through the abdominal wall into the amniotic fluid. Prostaglandins are naturally occurring hormones which in high concentration cause contractions. This kills the fetus. The woman must then go through an induced labour to deliver the dead fetus because it is too large to be removed by aspiration. The method is more distressing than the methods above.

■ **Hysterotomy (Caesarean)** – at 16+ weeks of pregnancy. This is only performed rarely because the physical risk to the mother is greater than with other methods. It is only used if other methods cannot be used or are unsuccessful. The fetus is removed through a cut in the abdominal wall and uterus, as with a Caesarean birth.

■ **RU486 (the 'abortion pill')** – at less than 10 weeks of pregnancy. RU486 is an anti-progesterone drug which results in the rejection of the embryo or fetus. The drug was introduced in

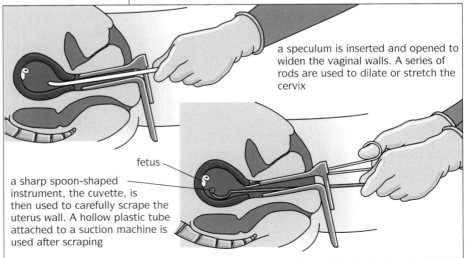

a speculum is inserted and opened to widen the vaginal walls. A series of rods are used to dilate or stretch the cervix

fetus

a sharp spoon-shaped instrument, the cuvette, is then used to carefully scrape the uterus wall. A hollow plastic tube attached to a suction machine is used after scraping

● **Figure 4.24** Dilation and curettage.

1988 and has been used extensively in France. Its long-term effects are unknown. It has now been licensed for use in Britain (few countries have yet done this). Three tablets are taken to start the procedure. Then one or two days later, a prostaglandin injection or pessary is given which causes contraction and completes termination. It is cheaper and easier than alternative abortion methods.

Using modern techniques, the physical health risk to the woman of having an abortion is less, on average, than that of completing the pregnancy. Day-care abortion is performed on an out-patient basis and usually enables a woman to return home three to four hours after the operation. The earlier an abortion is carried out, the less likely are medical complications. If the womb (uterus) or oviducts become infected, pelvic inflammatory disease, with possible infertility, is the most common complication. Other possible complications include the increased likelihood of ectopic pregnancy (the development of the fetus in the oviduct rather than in the uterus), a stretched and non-functional cervix, the perforation of the uterus, and retention of the placenta, leading to bleeding. An average of five women die each year from legal abortions in England and Wales and about 5% are made sterile by the operation.

Ethical issues

Not everyone is satisfied with the law as it stands. The National Abortion Campaign argues that women should have the right to choose whether to have an abortion at any time during pregnancy – 'abortion on demand' – and that all abortions should be freely available on the National Health Service. Such groups are often referred to as 'Pro-Choice' *(figure 4.25a)*. 'Pro-Life' groups, on the other hand, want the time limit lowered to 18 weeks, arguing, for example, that after 18 weeks a fetus feels pain. Recent evidence based on hormonal changes suggests that a fetus of 23 weeks can feel pain, but no-one knows for certain when the subjective experience of pain develops. In the USA, Pro-Life groups have become extremely militant and are a powerful political force *(figure 4.25b)*. Many Pro-Life groups are totally against abortion. Until July 1989 all women in the USA could have an abortion up to 12 weeks of pregnancy. However, changes in the law in the USA have given far more control to individual States, with some setting much stricter limits on abortion. By contrast, many countries, such as France and Denmark, allow abortion on request. In some countries, such as Russia, it is virtually the only form of birth control and the average woman has several abortions in her lifetime.

Being literally a matter of life and death, it is not surprising that strong views are commonly held. The ethical questions are

● *Figure 4.25*
a A pro-abortion rally
b An anti-abortion rally.

numerous. Some which are commonly asked are raised below, but full discussion here is not possible.

Some issues for discussion

■ Is abortion murder? Those of the Islamic and Christian faiths believe that the soul is independent of the body and is present from the moment of conception. The question might there-

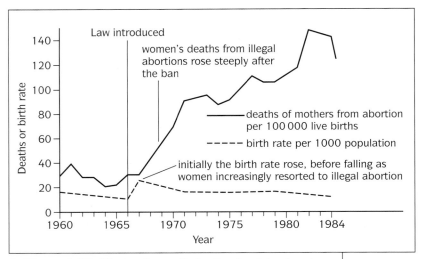

● **Figure 4.26** Effects of an anti-abortion law introduced in 1966 in Romania.

fore be easily answered by some. For others it is much more difficult. A common argument is that the fetus must reach a certain level of complexity before abortion should be regarded as murder. The ability to feel pain or to be conscious might be a suitable threshold. However, there is no general agreement as to when these events occur. The brain continues to develop in complexity throughout pregnancy.

■ Should the fetus have any rights? Should any humans have rights? You might like to discuss what we *mean* by rights and what rights we should have. Note that even if it is agreed that the fetus has no rights, this does not necessarily make abortion morally acceptable.

■ When does life begin? For a biologist there is no single moment or event which could be described as the creation of a new individual. It could be said to start with the growth of an egg, but life is cyclical since the egg must come from the pre-existing living cells of the mother. Some of the cells of the developing embryo form structures other than the fetus, such as the chorion, and until about 14 days after conception, the embryo can split into two or more individuals (identical twins, triplets, etc). Thus the cells which contribute to the uniqueness of an individual are not determined for some time after conception. Can ethical decisions be based on biology alone?

■ Should permission to carry out abortion be related to the ability of the fetus to survive independently outside the womb? The latter depends on the present limits of our technology. Should the right of the fetus to life depend on our technological development?

■ Should it be just the mother's right to decide? Is it valid to argue that, after all, it is her body? What about the father?

■ An estimated 2 million women die world-wide each year from illegal or unsafe abortions. A reduction in these abortions would result in fewer deaths and harmful side-effects on health *(figure 4.26)*. Also, the law is brought into disrepute when it is ignored on a large scale. Isn't it just as important to try to save the lives of women as their unborn babies?

■ What about a woman who has become pregnant after being raped?

■ Does using abortion to terminate handicapped fetuses lead to an uncaring attitude within society and the alienation of handicapped people? Or does it, as some believe, demonstrate a *caring* attitude?

■ If the mother does not want the baby, why doesn't she give it up for adoption at birth rather than have an abortion?

■ What about the health risks to the mother, both physical and psychological? Many women suffer feelings of guilt and regret afterwards and may need counselling. Some feel they will be punished in some way. These feelings can return after several years.

■ What about the existing family? An extra child may cause extra problems.

■ Many women are pressured into abortion. Should official attitudes to single mothers change? Should society make it easier for single parents to find work, guaranteeing crèche and nursery places?

■ What about evidence suggesting that unwanted children tend to do worse in education, employment and marriage?

- Should the drug RU486 be allowed when its long-term effects on health are not known?
- Since each abortion represents an *unwanted* pregnancy, should society atttempt to do more to promote successful birth control (see Contraception on page 53)? Are the levels of abortion acceptable? Should abortion be encouraged as a substitute for contraception in countries like Russia?
- In some countries, such as India and China, male children are preferred. Should sex-selection of offspring be allowed, and if so, is abortion an acceptable mechanism? There are moves to try to prevent this in India.

Perhaps the last word should be given to the International Planned Parenthood Federation (IPPF), an international voluntary organisation which has consistently campaigned for more effective contraception in order to *reduce* abortions. They argue that one of the causes of the high incidence of abortion is lack of knowledge of, and poor availability of, contraception. For example, a survey in Ghana revealed that only 13% of women having abortions had any knowledge of contraception. The IPPF regards freedom of reproductive choice as a basic human right and states that doctors should respect the wishes of those people seeking help in order to achieve the best interests of the patient.

In vitro fertilisation

In vitro fertilisation (IVF) is commonly known as the **test tube baby technique**. The first successful use of the technique came with the birth of Louise Brown in 1978. It was developed as a treatment for certain forms of infertility, mainly blocked or damaged oviducts. The term in vitro means, literally, 'in glass' and refers to the fact that fertilisation (fusion of the sperm with the egg) takes place *outside* the body. However, it usually takes place in a glass or plastic dish, and not a test tube. The technique can be used as a treatment for the following causes of infertility:

- Blocked or damaged oviducts. This can be a result of scar tissue forming in response to infection or abdominal surgery, or it may be an in-born error.

- Failure to produce eggs. Donor eggs can be fertilised and implanted.
- The male has very low sperm counts or poor sperm mobility, which will make normal intercourse unlikely to succeed. Fewer sperm are needed to fertilise eggs in vitro.
- The woman's cervical mucus is 'hostile', either killing or preventing the passage of sperm.
- The woman has been previously sterilised.

The technique may also be used if it is desirable to screen embryos for genetic defects before allowing pregnancy to proceed. An outline of the procedure is given below.

Preliminary tests and counselling are carried out to ensure suitability for treatment. Then from the beginning of the menstrual cycle, the woman being treated is injected daily with a hormone such as FSH which stimulates the growth of *many* eggs instead of the usual *one*. This increases the chances of success later. Progress in development of the egg follicles is followed by use of ultrasound scanning. Just before ovulation, the 'ripe' eggs are collected, most commonly by laparoscopy. This is performed under general anaesthetic. A narrow telescope is inserted through a tiny cut at the base of the navel into the abdominal cavity, which is inflated with carbon dioxide gas to make viewing easier. A fibre-optic cable connected to a microscope and TV monitor outside the body enables the surgeon to observe the internal organs. A long, fine, hollow needle is then inserted into the abdomen just above the pubic hairline, and the eggs are sucked from the follicles. The male partner produces a sperm sample and a special technique is used to collect the most mobile sperm from the sample.

The eggs are then 'matured' in an incubator for 4–24 hours in a special sterile culture medium. When they are mature, the sperm are added to the eggs and left for 24 hours for fertilisation to take place. The process of fertilisation can be assisted in a number of ways if sperm activity is low. The zona pellucida can be deliberately damaged to allow easier access to the egg, or, alternatively, the sperm can be micro-injected through the zona pellucida (a technique known as subzonal insemination, or SUZI), or into the cytoplasm of the egg. Two to

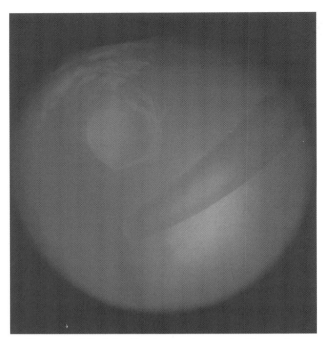

● *Figure 4.27* A fertilised ovum inserted with a soft tube into the uterus

three days later, any embryos that have formed are examined and a maximum of three suitable embryos are transferred into the uterus of the woman using a soft tube inserted through the vagina and cervix *(figure 4.27)*. The fate of any remaining embryos will already have been decided by the couple during preliminary counselling.

SAQ 4.14

Why are freshly ejaculated sperm incapable of fertilising an egg?

Ethical issues

The technique raises a number of important ethical issues. Probably the most important issue is the fate of unused embryos. These can be frozen and stored for future use, disposed of, or used for research,

depending on the wishes of the couple. There is some evidence that freezing mouse embryos may cause long-term behavioural changes in the animal. Disposal of human embryos could be regarded as murder (see Abortion on page 63). In the UK, research on embryos is allowed only up to 14 days. Up to this time twins can develop, and also the first rudiments of the nervous system appear. The research is valuable not just for improving infertility treatment for both male and female, but for developing techniques such as the diagnosis of genetic disease and gene manipulation. It can also be used to investigate reasons for miscarriage. It is worth noting that something like half of all human embryos are thought not to implant with natural pregnancies.

The success rate of IVF is only about 15–20%, but is rising. Even though couples are counselled beforehand, failure can be devastating, particularly after the lengthy treatment and many visits to the clinic. Very few National Health service clinics offer the treatment and private treatment is expensive. Apart from the obvious financial burden, particularly since more than one attempt may be necessary, there is the possibility that couples could be exploited in situations where the technique is even less likely to succeed than usual.

There is a risk of multiple pregnancy if more than one of the embryos introduced into the uterus develops. There is no evidence that children born as a result of IVF have a higher than average incidence of abnormalities.

The technique allows the possibility of embryo donation whereby the embryo is donated to another woman. This is basically an early form of adoption, whereby the adopting mother can give birth to the baby. Similarly, either eggs or sperm can be donated to other people. Such gamete donation raises ethical issues which are outside the scope of this book.

SUMMARY

■ The human reproductive system is closely linked with the excretory system in both sexes. The two systems are known collectively as the urinogenital system.

■ Female gametes, eggs, are produced in the ovaries; male gametes, sperm, are produced in the testes. Ovaries and testes are known as gonads.

■ Formation of gametes is known as gametogenesis. It involves meiosis of the nuclei of diploid mother cells. Sperm are formed continuously in the walls of the seminiferous tubules. Eggs are produced in cycles, one egg per month. The cycle is known as the menstrual cycle.

■ Gametogenesis is controlled by hormones. The control centre is the hypothalamus closely linked with the anterior lobe of the pituitary gland. The hypothalamus secretes GnRH (gonadotrophin releasing hormone) which stimulates release of gonadotrophins from the pituitary. These in turn stimulate the gonads.

■ The gonadotrophic hormones are FSH and LH (also called ICSH in the male). In response to these the gonads produce hormones and gametes. The male hormones are testosterone and inhibin; the female hormones are oestrogen and progesterone. These regulate gonadotrophin production, and thus gametogenesis, by negative feedback.

■ Fertilisation, the fusion of a sperm with an egg, occurs in the oviduct. Sperm are adapted to swim to the oviduct in the female after sexual intercourse, and to penetrate the follicle cells and zona pellucida around the egg. This penetration involves the acrosome reaction; capacitation of the sperm must take place before it is possible.

■ Human intervention into the normal reproductive process is common and increasing. Examples are contraception, abortion and in vitro fertilisation. All these processes involve ethical dilemmas.

■ Development of the fetus inside the mother is made possible by an organ unique to mammals, the placenta. It contains tissues from both mother and fetus and allows exchange of materials between the two.

■ Exchange of nutrients, respiratory gases, nitrogenous waste and some other substances is beneficial. Various transport mechanisms are involved, including diffusion, facilitated diffusion, active transport and pinocytosis. However, some harmful substances, such as certain drugs and viruses, can cross the placenta. It is therefore important that the mother is aware of how her actions can benefit or damage the fetus.

■ The placenta also acts as an endocrine organ, secreting chorionic gonadotrophin, oestrogen, progesterone and human placental lactogen. These help to maintain the uterus in a suitable state and stimulate breast development as well as having other important functions.

■ The fetus is suspended in amniotic fluid inside the amniotic cavity. It is thereby protected from physical shock and damage.

Questions

1 Explain how the **structure** of the placenta is adapted to its **functions**.

2 What advice would you give a pregnant woman concerning her diet?

3 Describe the changes that the following cells and their descendants must undergo before fertilisation is possible: **a** primary spermatocyte, **b** primary oocyte.

4 Some states in the USA have considered introducing laws to confine drug-dependent pregnant women to institutions during pregnancy to try to protect the unborn baby. What advantages and disadvantages can you think of for such legislation?

5 **a** What arguments could you use to try to persuade a woman to give up smoking during pregnancy?

   b Suggest ways in which the Government could encourage pregnant women not to smoke.

6 Discuss the advantages and disadvantages of increasing the availability of contraception worldwide.

7 **a** Contraceptives can prevent ovulation, fertilisation after ovulation, or implantation. Give examples of each type and explain the biological reason why contraception is achieved in each case.

   b Discuss the health implications of different forms of contraception.

8 **a** Describe ways in which hormones can be used by the medical profession to regulate and control human fertility.

   b Discuss the potential benefits and problems associated with the use of hormones in this way.

9 If you could change the law on abortion in the UK, what changes would you make, and why?

10 Explain how there can be economic, social and political aspects to the ethical issues associated with IVF.

11 Describe the trends shown by *tables AI* and *AII* (in the Appendix), including the trend in total numbers of abortions. It will help you to make some preliminary calculations. For example, for the years 1969, 1979 and 1989 in *table AII*, calculate the percentage of abortions that were carried out **a** under 9 weeks **b** at 9–12 weeks **c** at 13+ weeks of gestation. Carry out some appropriate calculations for *table AI*. Try to concentrate on the *main* trends. The trends could be discussed in class or in an essay.

12 Using the data provided in *table 4.5*, calculate the percentage of abortions carried out on single and married women in the years 1982, 1985, 1990 and 1992. Comment on the trend.

Year	Single	Married	Other (widowed, divorced, separated, unknown)
1982	71 836	40 510	16 207
1985	87 213	37 698	16 190
1990	116 150	38 151	19 599
1992	105 630	36 394	18 471

● **Table 4.5** Marital status of women having abortions in 1982, 1985, 1990 and 1992 (residents of England and Wales)

Control of growth and reproduction

1 explain the factors that control flowering in short-day and long-day plants;

2 describe the use of plant growth regulators in fruit maturation;

3 design and carry out an investigation to identify the major factors affecting germination;

4 describe the reasons for, and the advantages of, seed dormancy;

5 explain the role of plant growth substances in the control of seed dormancy;

6 describe the role of hormones in birth and lactation;

7 outline the role of hormones in premenstrual tension, the menopause and hormone replacement therapy;

8 outline the roles of the hypothalamus and pituitary gland in human growth and development;

9 describe the structure of the thyroid gland and the functions of thyroxine, the hormone it secretes;

10 describe the control of thyroxine secretion.

In chapter 1 we studied growth and in chapters 2, 3 and 4, reproduction. In this chapter we shall examine some of the ways in which the control of growth and reproduction is brought about in both flowering plants and in humans.

Genes, environment and coordination

Two major influences are at work in controlling growth and reproduction of living organisms: the genes and the environment. Important environmental factors include light, temperature, availability of nutrients and water. Ultimately, though, it is the genes, in the form of DNA, which have the blueprint for successful growth and reproduction. Complex internal coordination and control is required, both to express the DNA blueprint and to respond appropriately to the environment. Plants rely entirely on chemicals, whereas animals use chemicals, called hormones, *and* a nervous system. In this chapter we shall concentrate on the chemical control systems. In the first part of the chapter we shall focus on plants and use as examples flowering, fruit maturation, germination and dormancy. These processes provide excellent examples of how chemical control of growth and reproduction in plants is closely linked to, and influenced by, environmental change. In the second part, hormonal control in animals is illustrated by examples of the control of human growth and reproduction.

The control of flowering in plants

Plants rely entirely on chemicals for internal coordination of growth and development. The chemicals concerned are referred to as **plant growth substances** or **plant growth regulators**. They are sometimes referred to as hormones, but are not strictly comparable with animal hormones because they have different modes of action.

How does a plant know when to flower? You may well have assumed that it simply flowers when it has reached a certain stage of maturity, but this is not necessarily the case. It is often dependent on environmental conditions. Flowering is a very fundamental switch in activity. The flower parts are produced at shoot meristems where leaves and lateral buds are normally produced, so some fundamental change in gene expression must be brought about. The phenomenon is therefore of great interest to plant physiologists, and more recently to molecular biologists and geneticists as well. It has important commercial implications because plant growers would welcome the opportunity to be able to switch flowering on or off at will so that they can produce flowers and fruits out of season.

Photoperiodism

In 1910 it was discovered that the length of day (or to be more accurate, the length of time during which there is daylight) seemed to be involved in the flowering response. Moving away from the equator into temperate latitudes, the length of day shows increasing variation during the year. It is an almost constant 12 hours at the equator, but varies, for example, between 9 and 15 hours in the UK. Length of day is known as the **photoperiod** and the ability to respond to it is called **photoperiodism**. In 1920, N.W. Garner and H.A. Allard showed that tobacco plants would flower only after exposure to a series of short days. This occurs naturally in autumn in the UK, but can be induced in a greenhouse using artificial seven-hour days. Investigations of other plants revealed that some, such as spinach, would flower only in response to long days (**long-day plants** or **LDPs**). Some were unaffected by day length (**day-neutral plants**). Plants like tobacco were called **short-day plants** (**SDPs**). The other obvious environmental variable, temperature, was also investigated and was found to have a modifying effect on flowering in some cases. For example, some plants were day-neutral at one temperature and not at another.

The next important advance in understanding came in 1938 when K.C. Hamner and J. Bonner discovered that it was not day length that the plants were sensitive to! It was the length of the dark period. Thus SDPs ought strictly to be called long-night plants and LDPs should be called short-night plants, although the terms SDP and LDP are still used.

SAQ 5.1

Why might a link between flowering and day length be an advantage to a plant?

SAQ 5.2

What other environmental variable would be a useful cue for plants to flower?

SAQ 5.3

How could you prove that night length rather than day length is the critical factor?

(LNP)

If a SDP such as cocklebur (*Xanthium*) is given long nights, it flowers. However, Hamner and Bonner discovered that if the long night is interrupted by a short light period, flowering is prevented. Even an interruption of a few seconds can be long enough at high light intensities. Similarly, a LDP can be induced to flower in short days if the long night is interrupted by light.

Phytochrome and the action spectrum

Investigators then set about trying to discover which wavelength of light was responsible for reversing the effect of a long night, because this could help identify the pigment responsible for absorbing the light. *Figure 5.1* shows an action spectrum for preventing flowering in cocklebur. It was found that red light was most effective (with a wavelength of 620–660 nm); only two minutes' exposure was required to completely inhibit flowering. Later it was unexpectedly found that far-red light (730 nm) was effective in cancelling

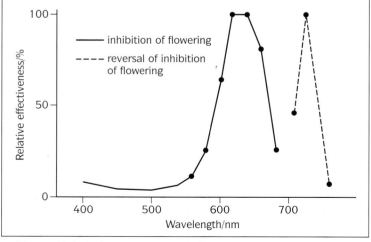

● *Figure 5.1* Action spectra for the effects of a light interruption of the long dark period on flowering in cocklebur.

out the effect of red light. In a series of light exposures, for example red, far-red, red, far-red, it was always the *final* stimulus which was effective. Flowering could be switched on and off.

This effect was later shown to be due to a pigment, named **phytochrome**. This was isolated in 1960 and shown to exist in two forms, a red-absorbing form, P_R or P_{660} (its absorption peak is 660 nm) and a far-red absorbing form, P_{FR} or P_{730} (absorption peak 730 nm). The absorption spectra of the two forms are shown in *figure 5.2*. Note how closely they match the action spectrum in *figure 5.1*. Phytochrome is a blue-green pigment and, as with haemoglobin, the pigment is attached to a protein. It is present in minute amounts throughout the plant, particularly in the growing tips, and it is involved in a range of plant responses to light, including greening of leaves.

Absorption of light by one form of phytochrome converts it rapidly and reversibly to the other form:

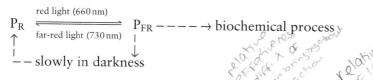

$$P_R \underset{\text{far-red light (730 nm)}}{\overset{\text{red light (660 nm)}}{\rightleftharpoons}} P_{FR} ----\rightarrow \text{biochemical process}$$

-- slowly in darkness

relative rates of effectiveness of diff. λ of light in bringing about a reaction

relative amounts of light of diff. λ absorbed by a chemical

SAQ 5.4

Summarise the difference between an **action spectrum** and an **absorption spectrum**.

Sunlight contains a lot more red light than far-red light. Therefore during the day, phytochrome exists in the P_{FR} form. At night it slowly changes back to the P_R form. This can be accelerated by using far-red light. In effect, the plant has a clock by which it can measure the length of the dark period; the longer the dark period, the more P_R. The conversion of P_R to P_{FR} in the day or in red light is more like a switch. The P_{FR} form is thought to be the physiologically active form and the P_R form is the inactive form. It seems likely therefore that P_{FR} stimulates flowering in LDPs (because short nights are not sufficient to remove all the P_{FR}). Conversely, it probably inhibits flowering in SDPs (because long nights remove the P_{FR} and hence the inhibition). Unfortunately for the validity of this theory, exposure to far-red light does not completely substitute for long nights, so some other, as yet unknown, time factor also appears to be important.

Role of growth regulators

It was shown in the 1930s that the light stimulus is perceived by the leaves and somehow transmitted to the flowering apex. This implies that a hormone may be involved. So convinced were plant physiologists that this hormone must exist that it was given a name, **florigen**. Despite intensive efforts, however, it has never been isolated and its existence is now in doubt. However, a class of growth substances called **gibberellins** can mimic the effect of red light in promoting flowering in certain LDPs, such as henbane, which are rosette plants (that is, they grow close to the ground). These plants must 'bolt', in other words the stem must elongate rapidly, before flowering. It may be that

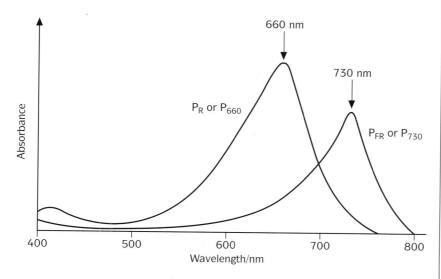

● *Figure 5.2* Absorption spectra of the two forms of phytochrome.

gibberellins simply promote bolting and that the subsequent flowering is controlled by another hormone. Gibberellins also inhibit flowering in some SDPs, providing further evidence that they may sometimes help to control flowering. Other plant growth regulators have been known to affect flowering in particular cases, but no overall pattern or theory of control has yet emerged.

SAQ 5.5

How could you demonstrate that the light stimulus for flowering is perceived by the leaves in a SDP?

Fruit maturation and ripening

Agriculture is the world's largest industry and fruits of various kinds are among the most important products. There is therefore a great deal of commercial interest in the process of fruit development and ripening. Usually, pollination, or pollination followed by fertilisation to produce seeds, is necessary to trigger the development of the fruit. Germinating pollen is a rich source of a class of plant growth substances known as the **auxins**. Pollination also seems to activate auxin production by the parts of the gynoecium such as the style and the ovules. Seeds are rich sources of two other groups of growth regulators, the **gibberellins** and **cytokinins**, both of which may be involved in fruit maturation.

Although developing seeds are usually essential for fruit growth, fertilisation does not always need to take place in order to trigger the development of the fruit. This has commercial significance because it is sometimes desirable to produce seedless fruit, a process known as **parthenocarpy.** It is common as a *natural* process among fruits which produce many ovules, such as figs, bananas, melons, pineapples and tomatoes. In 1939, F.G. Gustafson discovered that auxins could stimulate parthenocarpy in a number of plants, such as the tomato. Pollen grain extracts can sometimes do the same. This links well with the involvement of auxins in pollination noted above. It has also been shown that the horticulturally chosen varieties of bananas, pineapples, oranges and grapes which are

naturally parthenocarpic have unusually high levels of auxins compared with non-parthenocarpic fruits where, presumably, the seeds contribute the extra auxins needed for fruit development. Evidence for the suggestion that seeds contribute auxins is provided by the experiment illustrated in *figure 5.3*. Gibberellins will induce parthenocarpy in a few fruits which are unaffected by auxins. They are more effective than auxins at promoting parthenocarpy in tomatoes. A gibberellin is being used commercially to increase the size of seedless grapes, since naturally seedless grapes tend to be small and suitable mainly for raisin production rather than for eating. *Synthetic* auxins are used to improve fruit set and fruit size, for example in grapes. Fruit set is the retention of young fruit on the plant. In natural situations there is sometimes a

● **Figure 5.3** The control of flesh growth in the strawberry.
a A normal strawberry with the 'seeds' left on.
b When some of the 'seeds' are removed growth of the flesh occurs only under the 'seeds' which remain.
c Growth is stopped when all 'seeds' are removed.
d Growth is restored when the seedless strawberry is treated with auxins. Significant quantities of auxins can be extracted from the strawberry 'seeds'.
Note that the fleshy part of the strawberry is really the receptacle, and the true fruits are the pips. These may be regarded as seeds for physiological purposes.

heavy shedding of young fruit from the plant, with a consequent loss in yield. The commercial advantages of preventing this are great. Gibberellins are also used to increase fruit set, particularly in mandarins, tangerines and pears.

One of the difficulties in marketing fruit is trying to ensure that ripening takes place at just the right time to appeal to the customer. When you buy bananas, for example, you may well want the option of buying them slightly under-ripe or you may want them ready to eat. Days or weeks spent in transit therefore pose problems. What is needed is an understanding of the ripening process so that it can be artificially regulated.

As long ago as the 1930s it was known that the gas **ethene** speeds up the ripening of citrus fruits such as lemons. Later it was shown that ripe bananas and apples, and a whole range of other fruits, release ethene gas naturally. We now know that ethene is a natural growth substance which can probably be made by all plant organs. It is a simple organic molecule, with the formula C_2H_4. It is regarded as a **growth inhibitor** and stimulates the plant's ageing processes. It is particularly associated with leaf fall, fruit ripening and fruit fall.

Fruit ripening can be regarded as a senescence (ageing) process. It is typically accompanied by a burst of respiratory activity, which seems to be controlled by ethene. This process requires oxygen and can therefore be inhibited by lack of oxygen. Thus we have the principle by which fruit ripening can be controlled. Fruits can often be prevented from ripening by storage in an atmosphere lacking oxygen. Then, when ripening is required, for example at the end of a long sea voyage, it can be stimulated by adding oxygen and ethene. This method is applied commercially to bananas and citrus fruits. Ethene is also used to stimulate ripening of tomatoes.

Factors affecting germination and dormancy

Germination is the onset of growth of the embryo in a seed, usually after a period of dormancy. **Dormancy** is the state where germination will not occur, even if environmental conditions are favourable. It is sometimes controlled by plant growth substances. The effects of plant growth substances and the environment on germination can be investigated experimentally. *Boxes 5A, 5B and 5C* describe investigations that can be carried out into the roles of environmental factors and growth regulators on germination and dormancy.

Box 5A An investigation of environmental factors affecting germination

Certain environmental conditions are required for germination, including the presence of water and oxygen, and the correct temperature. *Figure 5.4* shows a range of experimental test tubes which can be used to demonstrate the need for each of these, using mustard or cress seeds. These are small enough to germinate in test tubes on a support of cotton wool.

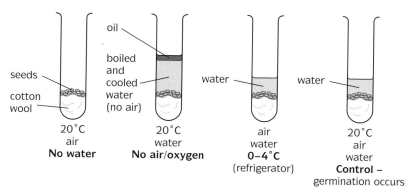

Replication: Use more than 1 seed per tube. 30 shared between several tubes for each condition should be ideal.

Quantify: More quantitative results could be obtained by calculating % germination. A range of temperatures could be attempted, if incubator facilities are available, and the optimum temperature found graphically.

● **Figure 5.4** Experiment to investigate environmental factors affecting germination of cress seeds.

Box 5B An investigation of the effects of abscisic acid, kinetin and light on the breaking of dormancy

Abscisic acid (**ABA**) is a natural growth inhibitor which interacts with other growth substances to regulate growth.
Kinetin is a synthetic cytokinin, one of the three groups of growth substances in plants which stimulate growth.

Some seeds, such as some varieties of lettuce, need light for germination. This is a phytochrome-controlled response and red light is most effective. Using seeds of a light-sensitive variety of lettuce, such as Dandie or Kloek, or simply a range of seeds, and solutions of ABA and kinetin of appropriate concentration, an experiment can be set up to investigate the effects of ABA, kinetin and white light on germination. Most seeds can be conveniently germinated on filter paper inside petri dishes.

The following experimental dishes could be set up:

A ABA present
B kinetin present
C kinetin + ABA present
D water (as a control)

These could be left in white light. A second set of 4 dishes could be set up and left in the dark. Addition of 50 seeds per dish would be appropriate, after soaking the filter papers in 5 cm^3 of the appropriate solution and removing any air bubbles. The dishes should be incubated at 25°C and percentage germination recorded in each dish after 2–4 days. Final concentrations of growth substances in the dishes should be ABA 1 ppm (parts per million) (1 g per million g water, or 1 mg per dm^3), and kinetin 10 ppm. A convenient way of doing this is to prepare stock solutions of ABA 2 ppm and kinetin 20 ppm and to add the appropriate solution or water to the dishes as follows:

A 2.5 cm^3 ABA + 2.5 cm^3 water
B 2.5 cm^3 kinetin + 2.5 cm^3 water
C 2.5 cm^3 kinetin + 2.5 cm^3 ABA
D 5 cm^3 water

Box 5C An investigation of the role of gibberellins in breaking dormancy of cereal seeds

As we have seen, gibberellins are one of the three classes of growth substance that promote growth. One of their effects is to break the dormancy of cereal seeds. This is achieved by stimulation of enzymes which hydrolyse (digest) the food reserves in the endosperm. In particular, activity of the enzyme α-amylase is stimulated, which catalyses starch digestion.

SAQ 5.6

The following experiment was set up to demonstrate production of α-amylase by germinating barley. Five barley grains were cut in half so that in each case one half contained the embryo and the other just endosperm. The 'non-embryo' halves were discarded. The 'embryo halves' were surface sterilised (bacterial and fungal contaminants can also produce α-amylase) and placed in a petri dish containing sterilised starch-agar medium (agar jelly containing starch). They were incubated at 25°C for 48 hours. The surface of the starch agar was then flooded with iodine solution. The final appearance of the dish is shown in *figure 5.5*. The 'halos' around each grain are due to the digestion of starch by α-amylase which has diffused out of the grain. In the rest of the agar, iodine solution has reacted with starch to give a blue-black colour. How could you gain evidence to support the hypothesis that gibberellin

a stimulates α-amylase production in germinating barley grains, and **b** is produced in the embryo.

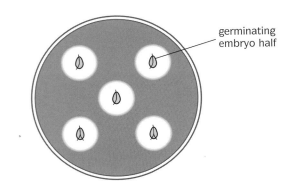

● *Figure 5.5* Final appearance of petri dishes.

Seed dormancy

It is often an advantage for a seed to undergo a period of dormancy. Germination can then be linked to important factors such as a favourable season. For example, a seed released in autumn is more likely to germinate successfully if it remains dormant throughout winter, even if conditions for growth are favourable in autumn. Some of the common mechanisms for controlling dormancy are discussed below.

Growth inhibitors

Some seeds, such as ash, contain growth **inhibitors**. A wide variety of chemicals act as inhibitors, from sodium chloride (high concentrations of which cause osmotic inhibition in some seeds) to complex organic compounds such as essential oils and alkaloids. Abscisic acid (ABA) is the most common inhibitor. The inhibitors may be lost gradually by being leached out by water, or their effects may be overcome by growth promoters such as gibberellins. Many fruits, including the tomato, contain inhibitors which prevent the seeds germinating while still inside the fruit. Indeed, tomato juice is a potent growth inhibitor of the seeds of many species.

SAQ 5.7

Why would the need for leaching of inhibitors from its seeds be an appropriate control mechanism for the germination of a desert plant?

Temperature

Many seeds require an exposure to low temperatures in moist conditions with oxygen for several weeks or months, a process known as **prechilling**. Cereals, peach, plum, cherry and apple are examples *(figure 5.6)*. This prevents the seeds from germinating in autumn or during a warm spell in winter. The mechanism is still unclear and probably varies from species to

species. The latest evidence suggests that the embryo is the part of the seed that is sensitive to the low-temperature stimulus. Adding gibberellins sometimes substitutes for prechilling, so perhaps prechilling brings about a natural increase in gibberellins in some species.

Light

Light is an important factor in controlling the germination of the seeds of many wild plants. (Humans have tended to select out this response from crop plants.) Light *stimulates* the germination of many seeds, for instance lilac and some varieties of lettuce (see *box 5B* on page 76). On the other hand, the germination of the seeds of some other species is *inhibited* by light.

SAQ 5.8

Suggest a reason for the fact that light-stimulated seeds are often small and without large food reserves.

SAQ 5.9

Suggest an advantage for a seed whose germination is **inhibited** by light.

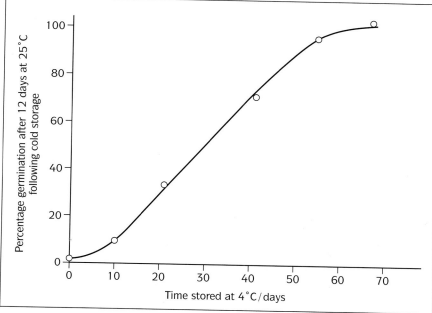

● **Figure 5.6** Germination of apple seeds after different periods of storage at 4 °C.

SAQ 5.10

SAQ 5.10

When light passes through leaves, most of the red and blue wavelengths are absorbed, but most of the far-red light passes through. Far-red light often inhibits germination of seeds that require normal sunlight for germination. Suggest an ecological advantage for this.

SAQ 5.11

Which pigment is responsible for the far-red inhibiting effect?

Scarification

Some seeds require physical damage to the seed coat (the testa) before germination. Such damage is called **scarification**. Without it, the testa may be tough enough to prevent growth, or may prevent oxygen or water uptake. This can be demonstrated by artificially scarifying the seeds; sandpaper and knives have been used, but sometimes even a pin prick is sufficient. The phenomenon is ecologically important. For example, certain species can be stimulated to germinate by exposure to fire. The seeds of one species of *Albizzia*, an Australian tree of the legume (pea and bean) family, contain a small plug in the seed coat which pops out when the seed is heated. This allows entry of water for germination. In areas where fire is common, such as Australia and southern California, scarification by fire ensures early recovery of vegetation after fires. Other natural ways in which scarification may be achieved include gradual attack by bacteria, or passage through the gut of a bird or other animal. The latter case also helps the dispersal of the seeds, with the added advantage that the egested waste in which the seeds are deposited is a ready-made fertiliser.

The physiology of germination

Having looked at environmental and hormonal influences on dormancy and germination, we can now summarise how these factors interact. In *box 5A* it was established that water, oxygen and a suitable temperature are required for germination.

Germination is initiated by the uptake of water, a process known as **imbibition**. It takes place through the micropyle and testa. As the contents of the seed hydrate, they swell and rupture the testa (and pericarp if necessary), allowing the radicle and plumule to emerge. Water has several vital functions. It is an essential solvent, allowing biochemical reactions to take place in solution, and the transport of nutrients from the food reserves to the growing embryo. It is also a reagent, taking part in the many hydrolysis (digestive) reactions by which the food stored in the cotyledons or the endosperm is broken down. Overall, therefore, water activates the seed.

Oxygen is needed for aerobic respiration. The respiratory substrates, mainly glucose and lipids (usually oils), come from the food reserves. The resulting release of carbon dioxide is responsible for the initial loss in total dry mass of the seed described in chapter 1 (page 8 and *figure 1.10*). If there is an oxygen shortage, seeds can also respire anaerobically, producing ethanol and carbon dioxide from glucose.

Since the biochemical reactions of germination are controlled by enzymes, temperature affects the rate at which they act. There will be an optimum temperature for germination and a range outside which germination will not occur.

During the early phases of germination, mobilisation of the food reserves takes place. This is controlled by hormones such as gibberellins, abscisic acid, cytokinins, and auxins, as investigated in *boxes 5B* and *5C*. The role of gibberellins in cereal seeds is a good example (*box 5C* and *SAQ 5.6*). Here gibberellins are produced by the embryo at the onset of germination and stimulate the synthesis of new, key enzymes. This involves switching on the relevant genes which code for the enzymes. Similar events occur in all germinating seeds. One role for the enzymes is to hydrolyse (digest) the food reserves. Carbohydrates (such as starch), lipids and proteins are hydrolysed to sugars, fatty acids and glycerol, and amino acids, respectively. These products are used by the growing embryo, together with the minerals and vitamins also stored, as building blocks for new biochemicals, or as respiratory substrates. For example, as cells divide, new cell walls must be made, using glucose to make cellulose – as described in *Foundation Biology* on page 38. Amino acids are needed to synthesise new proteins, including enzymes and structural proteins. In the embryo, cell division takes place in the apical meristem of the root (the radicle) and the apical meristem of the shoot (the plumule). Zones of elongation and differentiation occur behind the meristem as described in chapter 1 on page 2.

Hormonal control in animals

As noted at the beginning of this chapter, the chemical control of growth, development and reproduction in animals is brought about by hormones. Hormones have been studied in chapter 6 of *Foundation Biology* (pages 67–9 and 73–4 are particularly relevant). It would be useful to check your knowledge by trying to answer SAQ 5.12.

SAQ 5.12

a Why are hormones referred to as chemical messengers? **b** Hormones only affect specific targets; how is this specificity achieved? **c** What characteristics do hormones share? **d** What characteristics are shared by all endocrine glands?

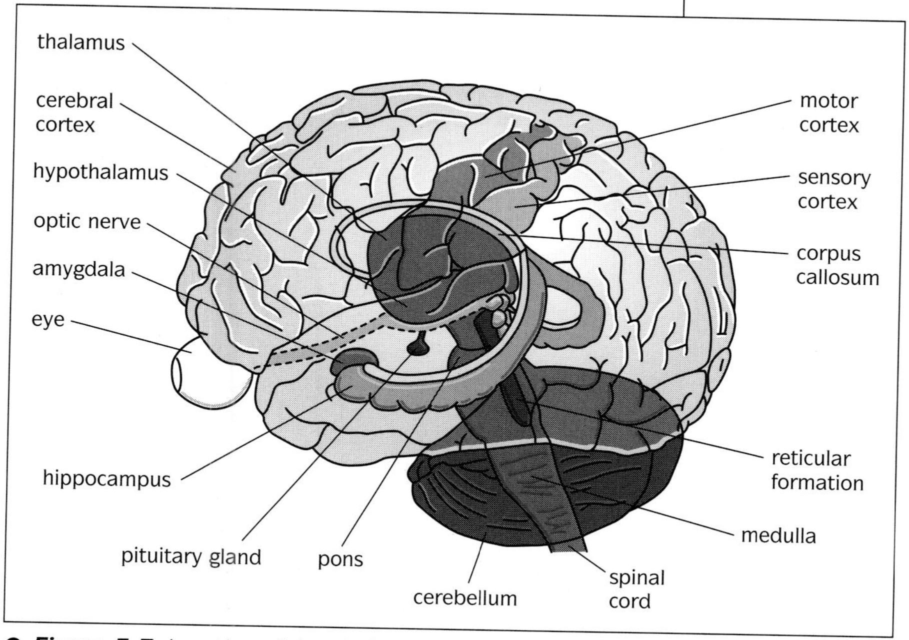

The hypothalamus and pituitary gland

In humans, all aspects of sexual reproduction, including puberty, pregnancy, birth, lactation and menopause are orchestrated by hormones. We have already seen that the hypothalamus and pituitary gland are the major control centres for these hormones (*figures 5.7, 4.13, 4.14* and *4.15*). This is also true for growth and development. The hypothalamus is a link between the endocrine system and the nervous system. It is a relatively small though complicated structure lying at the base of the brain. It receives information from other parts of the brain and from the blood flowing through it and regulates the activity of the pituitary gland, an endocrine gland which in turn regulates other endocrine glands. Hence the pituitary gland is sometimes known as the 'master gland'. It is about the size of a pea and hangs down from the hypothalamus on a short stalk (*figure 5.7*). It has two main parts, the anterior and posterior lobes. Each lobe releases its own particular hormones. Many feedback mechanisms operate on the hypothalamus and pituitary. Environmental factors such as seasonal changes can have an influence, as can psychological and physical factors such as emotions, stress and exercise. We shall focus on some of the better understood examples of control and coordination.

Role of hormones in reproduction

The vital role that hormones play in helping to control reproduction has already been seen in chapter 4, where spermatogenesis, the menstrual cycle and pregnancy were discussed. In this chapter we shall look at further ways in which hormones are involved in reproduction.

● *Figure 5.7* Location of the pituitary gland and hypothalamus in the brain.

Role of hormones in birth

Towards the end of pregnancy an as yet unknown signal, probably from the fetus itself, triggers the events which lead to labour and birth. The muscle layer of the uterus becomes increasingly sensitive to the hormone **oxytocin**, due to a decline in progesterone levels. Oxytocin is made in the hypothalamus and passes down special nerve cells to be released from the posterior lobe of the pituitary gland. Its target is the uterus, where it causes muscle contractions. There now occurs a rare example of **positive feedback** (most feedback mechanisms are negative). Uterine contractions stimulate release of more oxytocin via the nervous system and hypothalamus. This in turn stimulates more contractions. Thus during labour, contractions get more frequent and more powerful. The contractions force the fetus through the cervix and into the vagina from where it emerges to the outside world. Further contractions result in the delivery of the placenta or 'afterbirth'. To help this along, the midwife often injects some more oxytocin into the placenta after the baby is born.

Role of hormones in lactation

Lactation is the production of milk by the breasts. The breasts grow and develop during pregnancy (on average doubling in weight). Oestrogen and progesterone help to control this development, as shown in *figure 4.23*. They can only have this effect in the presence of **human placental lactogen (HPL)**. Milk, however, can only be produced in the presence of **prolactin**, a polypeptide hormone secreted by the anterior lobe of the pituitary gland in response to **prolactin releasing factor** from the hypothalamus. Prolactin levels build up during pregnancy, but the presence of high levels of oestrogen and progesterone, secreted by the placenta, restricts its release and therefore prevents milk production. At birth, loss of the placenta results in a lot more prolactin being released by the pituitary. Its target is the breasts, where it stimulates the glandular cells that produce milk. These cells are surrounded by contractile tissue, which squeezes the milk into ducts leading to the nipples. **Oxytocin** stimulates contraction of this tissue. The sucking of the baby on the breast sets up a nervous reflex which stimulates the hypothalamus to release more oxytocin and maintain milk flow. Thus, oxytocin stimulates milk ejection and prolactin stimulates milk production. Both hormones depend on suckling for their continued synthesis.

SAQ 5.13

Describe the broad overall changes in concentration that you would expect for the following hormones in the mother's blood during pregnancy and lactation: **a** luteal progesterone; **b** placental progesterone; **c** prolactin; **d** oxytocin; **e** HPL; **f** HCG.

Role of hormones in premenstrual tension

The term **premenstrual tension (PMT)** was first used in 1931 to describe the 'distressing psychological or physical symptoms' which some women experience regularly towards the end of each menstrual cycle and which 'significantly regress throughout the rest of the cycle'. Tension is not the only symptom. In fact, more than 150 symptoms have been attributed to PMT at various times, and the term **premenstrual syndrome (PMS)** was introduced in 1953 to acknowledge this variety of effects. Great controversy now surrounds the condition. In 1993, certain psychologists (not all of them men) went so far as to suggest that it may not exist at all, that it was more of a 'social construct' than a medical condition, and that it was a way of 'legitimising and expressing distress'. However, some women who experience it describe the condition as devastating and about 75% of women are said to be affected in some way. The most common symptoms are depression, changes in mood, water retention and aches and pains. Little is known about the cause, but it is usually assumed to be hormonal in origin. Since it takes place in the few days before menstruation, it could be due to changes in the balance between progesterone and oestrogen (*figure 4.15*), which decline at different rates at this time. Dr Katharine Dalton, who coined the term PMS, believes it may commonly be due to progesterone deficiency.

Menopause and hormone replacement therapy

Menopause is the cessation of monthly periods and marks the end of a woman's fertility. The average age of menopause in the UK is 51 years. Periods usually become irregular, before finally ceasing. The cause is the gradual failure of the ovaries. The number of follicles declines and they become less sensitive to FSH so that eggs are less and less likely to be produced each month. Secretion of oestrogen declines and since oestrogen normally inhibits FSH by negative feedback, higher levels of FSH (and later LH) are typical of menopause. Many symptoms, both physical and psychological, are associated with menopause and these are mostly due to the reduced oestrogen levels (although progesterone levels also decline). The commonest symptoms are night sweats, random hot flushes during the day, and vaginal dryness. Other common symptoms are depression, irritability, fatigue, and softening of the bones due to loss of minerals, particularly calcium. Loss of calcium from the bones causes a condition known as **osteoporosis**. It is characterised by loss of bone mass; as a result the bones break more easily. It occurs because oestrogen is antagonistic to the hormone **parathormone**, which stimulates the raising of blood calcium levels. The symptoms can be prevented relatively easily by **hormone replacement therapy (HRT)**, in which oestrogen is taken either in pill form or by implants below the skin. HRT greatly reduces the rate at which calcium is lost from the bones, slowing it down to roughly the same rate as in men. Treatment can be short-term or continued for years, although in the long term, blood clotting and other undesirable side effects may occur. Some of these can be prevented by adding progesterone to the oestrogen.

Role of hormones in growth and development

Growth hormone

The anterior lobe of the pituitary gland produces a hormone called **growth hormone (GH)**. It has no specific target organ , but regulates the growth of all parts of the body. It does this by stimulating protein synthesis. It is particularly important for development of limb bones. Thus an excess of growth hormone results in gigantism and a deficiency results in dwarfism, although brain development and IQ are unaffected. GH also favours the use of fat rather than carbohydrate for energy so the body becomes less fat and more muscular. Secretion of GH itself is under the control of the hypothalamus, with both a stimulating hormone and an inhibitory hormone being involved. Feedback inhibition from GH appears not to occur.

Secondary sexual characteristics

At the beginning of puberty, the hypothalamus begins to release GnRH, which stimulates the secretion of FSH and LH from the pituitary gland. These in turn stimulate the testes to produce testosterone, or the ovaries to produce oestrogens. These hormones are responsible for the development of **secondary sexual characteristics** such as increased body hair and enlarged genital organs.

The thyroid gland

The thyroid gland secretes two hormones which influence growth and development by affecting the rate of metabolic activity. These hormones are **thyroxine** (or T_4) and **tri-iodothyronine** (or T_3). The numbers refer to the number of iodine atoms per molecule. About 90% of the secretion is of thyroxine, and the two hormones act in a similar way, so we shall only consider thyroxine in the following account.

The thyroid gland has two lobes and is arranged like a bow tie around the front of the trachea in the neck. *Figure 5.8* shows the structure of the gland. It is made up of many follicles, each of which is a hollow sphere surrounded by a single layer of secretory cells (epithelium). These cells secrete a glycoprotein called **thyroglobulin**, which accumulates in the follicles. Thyroglobulin is a large molecule, containing iodine.

When thyroxine is to be released into the bloodstream, the

secretory cells take up small amounts of the stored thyroglobulin from the follicles by pinocytosis. Inside the pinocytotic vesicles, proteinases hydrolyse the large thyroglobulin molecules into smaller thyroxine molecules. These then travel through the cell, and pass into the blood capillaries.

Thyroxine is a small molecule carried in solution in the blood plasma, attached to plasma proteins. These bind quite tightly to the hormones, only releasing them slowly to body cells. Once released, the thyroxine enters body cells.

Role of thyroxine in growth and development

Thyroxine controls **basal metabolic rate (BMR)**. Metabolism is the collective name for all the chemical processes going on in the body. The basal rate is the rate at rest. It is the rate at which oxygen and food are used to release energy, and is directly related to the rate of cell respiration. When

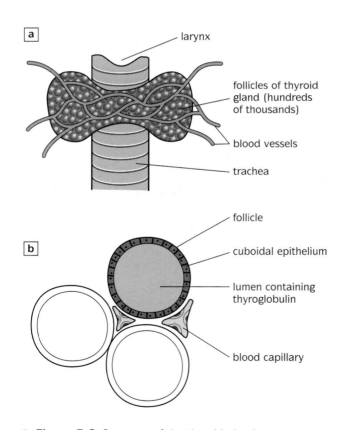

● **Figure 5.8** Structure of the thyroid gland.
a Position in the neck
b Transverse section showing three follicles.

thyroxine enters cells, it binds to specific protein receptor molecules in the cell nucleus. The combined hormone–receptor molecule attaches to specific regions of DNA and 'switches on' transcription of mRNA, thus bringing about synthesis of whatever protein is coded for by that particular piece of DNA. Many different genes are transcribed, leading to the production of a wide variety of enzymes. This increases the metabolic rate of the cell. The rate of respiration in particular increases.

Thyroxine influences growth and development in a number of ways. It stimulates growth in general, but particularly protein synthesis and development of the skeletal system. Unlike growth hormone (GH), it also stimulates brain development, so a deficiency in children can result not only in dwarfism, but mental retardation and low IQ (cretinism). It has a number of other effects, including increasing heart rate and cardiac output.

The hormone is released only slowly from plasma proteins over days or weeks. Once it has entered its target cells, it gradually switches on genes, and so the basal metabolic rate only changes very gradually over weeks or maybe even months.

Control of thyroxine secretion

The control of thyroxine secretion is summarised in *figure 5.9*. A straightforward negative feedback system operates, again involving the pituitary gland and hypothalamus, as well as two other hormones. The pituitary gland switches on the thyroid gland by secreting **thyrotrophin releasing hormone (TRH)**. As well as the negative feedback system, higher centres of the brain may also be involved, which may in turn respond to environmental cues. For example, the basal metabolic rate may be increased during cold seasons to help maintain body temperature.

The cells of the pituitary gland are sensitive to the amount of thyroxine in the blood passing by them. If thyroxine levels are low, the anterior pituitary gland secretes **thyroid stimulating hormone (TSH)**, into the blood. TSH is carried in the blood to the thyroid gland, where it stimulates the secretion of thyroxine. High levels of thyroxine in the blood switch off the secretion of TSH.

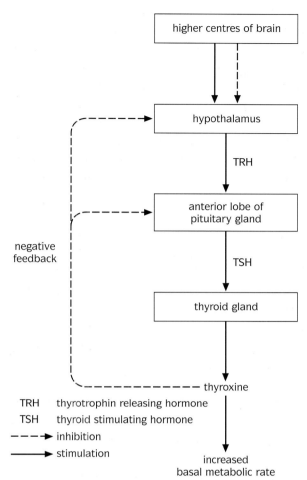

TRH — thyrotrophin releasing hormone
TSH — thyroid stimulating hormone
----▶ inhibition
——▶ stimulation

● *Figure 5.9* Control of thyroxine secretion.

SUMMARY

■ Internal coordination and the control of growth and development are genetically programmed. In plants they depend entirely on chemicals. These chemicals are referred to as growth regulators or growth substances. In animals both chemicals and a nervous system are involved and the chemicals are referred to as hormones.

■ In both plants and animals the environment also has an important influence on growth and development. These processes can therefore be seen to be a product of the interaction between genes and the environment.

■ There are five classes of plant growth substance, namely auxins, gibberellins and cytokinins, which are growth promoters, and ethene and abscisic acid, which are regarded as inhibitors.

■ Flowering, fruit maturation, germination and dormancy are good examples of the interaction between environment and growth substances in plants.

■ Flowering is often controlled by day length (photoperiod). In such cases, some plants flower in response to short days (short-day plants) and some in response to long days (long-day plants). The blue-green pigment phytochrome is responsible for this. It is converted to an active form by absorption of red light; far-red light or an extended dark period reverses this process. A hormone that controls flowering has not been found.

■ All the growth promoters are involved in fruit growth, particularly auxins. Fruit ripening is often stimulated by ethene.

■ Germination requires particular environmental conditions, including water, oxygen and a suitable temperature. It can only occur once dormancy is broken.

■ Dormancy is often caused by growth inhibitors, and broken by growth promoters. Gibberellins break dormancy in cereal seeds.

■ The process of labour and birth is stimulated by oxytocin, a hormone produced by the hypothalamus and released from the posterior lobe of the pituitary. Oxytocin stimulates contraction of muscle in the uterus wall. Its secretion is regulated by positive feedback.

■ Premenstrual tension is probably hormonal in origin and may be due to changes in the balance between oestrogen and progesterone towards the end of the menstrual cycle.

■ The menopause is caused by failing ovaries, which leads to a decline in oestrogen levels. The undesirable consequences, such as a greater risk of osteoporosis, can be avoided by hormone replacement therapy. This involves giving oestrogen pills or implants.

■ The hypothalamus and pituitary gland influence growth and development as well as reproduction. The anterior lobe of the pituitary, under the influence of the hypothalamus, secretes growth hormone which stimulates protein synthesis and the development of limb bones.

■ Development of secondary sexual characteristics is triggered once the hypothalamus releases gonadotrophin releasing hormone (GnRH).

■ The thyroid gland secretes thyroxine, which controls basal metabolic rate and thereby influences growth and development. Limb growth and brain development are particularly affected.

■ Secretion of thyroxine is stimulated by TSH (thyroid stimulating hormone) from the pituitary, which in turn is stimulated by TRH (thyrotrophin releasing hormone) from the hypothalamus. Thyroxine regulates its own production by negative feedback.

Appendix: Further data on abortion

Year	Women with no previous children	Women with one child	Women with two children	Women with three children	Women with four children	Women with five children and over	Not stated	Total
1969	21389	5345	8200	6514	3874	4572	448	49829
1970	33196	8204	12830	9980	5809	5192	751	75962
1971	43413	10383	16307	12092	6506	5269	600	94570
1972	50645	12581	19357	13710	6919	5157	196	108565
1973	52865	12990	20370	13324	6382	4467	170	110568
1975	52733	13119	19930	11789	5257	3168	223	106224
1977	52864	12799	19344	10494	4204	2372	600	102677
1979	64400	15496	22855	11174	4104	2086	496	120611
1981	67400	16746	23498	11050	3882	–	4026	128581
1983	69979	17030	22412	10109	3475	1755	2615	127375
1985	82964	19199	23103	9977	3279	1494	1085	141101
1987	92256	22158	24739	10301	3374	1626	1737	156191
1988	100644	24616	26005	10931	3480	1663	959	168298
1989	100543	25516	26713	11184	3573	1752	1182	170463
1990	100781	26846	27506	11429	3746	1832	1760	173900
1991	94523	27328	27780	11503	3831	1898	513	167376
1992	87731	27418	27504	11564	3975	2013	290	160495

1969–1979 live births only; 1981–1992 live or stillbirths

1969–1979 figures are based on notifications received. From 1981 onwards the figures are based on occurrences

1968 figures have been omitted since abortions were only legal from April 1968

● **Table AI** Abortions by number of previous births for residents of England and Wales between 1969 and 1992

Year	Under 9 weeks	9–12 weeks	13–19 weeks	20+ weeks	Not stated
1969	6644	25552	15629	1172	832
1971	15700	54747	20583	847	2693
1973	24053	64171	18513	975	2856
1975	25028	59615	16814	971	3796
1977	25227	57504	15199	1095	3652
1979	28882	67134	18622	1767	4206
1981	39819	68653	17379	1715	1015
1983	43890	64971	16671	1748	95
1985	48080	74172	16709	2116	24
1987	53821	81961	18176	2222	11
1989	60315	89450	23633	2817	2
1991	58873	88621	–	–	–
1993	61913	78489	–	–	–

● **Table AII** Abortions according to weeks of pregnancy at the time of abortion for residents of England and Wales between 1969 and 1993 (comparable data for the last three columns unavailable for 1991 and 1993)

Answers to self-assessment questions

Chapter 1

1.1 Mitosis

1.2 Different genes are switched on and off in different cells. Genes control cell function.

1.3 Growth could be measured as an increase in: size (e.g. height, length, volume); dry mass or fresh mass; number of cells; complexity.

1.4 **a** A = rate of growth (rate declines after initial rapid increase); B = dry mass (could not be zero at Day 0 and could not be negative up to 11 days).

b Shoot length would show no initial decrease.

c A = absolute growth rate curve;
B = absolute growth curve.

d Sigmoid.

e Respiration results in the breakdown of materials to generate energy and will therefore result in a loss of mass whereas photosynthesis is the synthesis of materials and results in an increase in mass. Immediately after germination the rate of respiration exceeds rate of photosynthesis. At first there is *no* photosynthesis. Later, it takes time for the rate of photosynthesis to increase to the point where the rate of increase in mass due to photosynthesis exceeds the losses due to respiration.

1.5 Advantages:

- measurement of mass is more representative than measurement of single dimensions such as height or length;
- fluctuations in water content have no effect on measurements, so dry mass is a better measure of *permanent* increase in size.

Disadvantages:

- Dry mass may decline when seeds germinate, until the seedling starts to photosynthesise;
- dry mass is more time consuming and difficult to measure than fresh mass;
- determination of dry mass is destructive;
- dry mass may not change very much when nutrients are diverted from one place to another, as in seed germination and seedling growth before photosynthesis, or during seed and fruit development when nutrients are diverted from leaves.

1.6 Increase in fresh mass. Increase in size of an anatomical feature, e.g. length of hind leg, length of developing wing, length of head. Mass of food eaten.

1.7 Head - bears sensory structures/feeding structures. These are important throughout life and important they are fully developed early.

Tibia - related to growth of leg – must increase in proportion to rest of body for efficient locomotion.

Wings – not needed until adulthood (for migration to new food sources) and expanded wings cannot moult.

Chapter 2

2.1 If gametes did not have half the number of chromosomes of normal body cells, the number of chromosomes would double with each generation as a result of the fusion of male and female gametes.

2.2 a Genetic uniformity and hence preservation of good characteristics; bulbs have larger food reserves than seeds and hence are more likely to survive; bulbs get off to a quicker start than seeds because the young plant is better developed.

b Lack of the genetic variation found among seeds means that better variants cannot be selected; it may be cheaper or less labour intensive to sow seeds rather than bulbs.

Chapter 3

3.1

Characteristic	White deadnettle (insect pollinated)	Meadow fescue (wind pollinated)
petals	large, conspicuous nectaries present	no petals no nectaries no scent
	landing platform for bees	no landing platform
stamens	inside flower stamens and anthers do not swing freely	hang outside flower to catch wind stamens and anthers swing freely in air currents
pollen	rough surfaced	light, small, smooth surfaced relatively large amount produced
stigma	inside flower sticky and lobed to trap pollen	feathery and hangs outside flower, therefore traps pollen more easily

3.2 a The gametes produced by one individual show genetic variation as a result of meiosis. **b** All the diploid cells of a given plant are genetically identical since they are all derived from the original zygote by mitosis. This includes the tissues of the flowers. Genetic variation only arises once meiosis takes place. Thus, just as much variation can exist among the gametes produced by one flower as between gametes from different flowers.

3.3 a Stigma above anthers = pin-eyed; stigma below anthers = thrum-eyed.

b Pollen deposited on the body of the bee while it feeds in the tube of one flower tends to pass to the stigma at the same level in another flower.

3.4 The *tube nucleus* controls growth of the pollen tube. The *generative nucleus* divides by mitosis to form the two male gametes.

3.5 It is an adaptation to life on land. Fertilisation is dependent on water if swimming sperm are released from the plant. Pollen grains, with their protective waterproof outer walls, are ideally suited for transport in dry conditions.

3.6 Vacuoles almost disappear since these are the sites of most water within plant cells. Food stores increase, for example oils and starch.

3.7 During meiosis of a triploid nucleus, homo-logous chromosomes come together in threes (trivalents) on the spindle rather than in pairs as in diploid cells. Organised separation of the chromosomes to opposite ends of the spindle is impossible since two of each trivalent must go one way, and one the other.

The endosperm nuclei never divide by meiosis. No other cells are derived from the endosperm. Ultimately they all die within the seed, being used only as a source of food.

Chapter 4

4.1 **a** **(i)** 4 **(ii)** 2 **b** spermatid

4.2 Similarity: haploid.

Differences:

Egg	Sperm
relatively large – stores nutrients to survive early stages of development after fertilisation (it also divides into smaller cells at this stage)	relatively small – economy of material and energy
contains food store	no food store – short-lived
stationary	motile – must travel to reach egg
only one produced at a time – multiple pregnancy avoided (usually)	millions produced – large wastage
produced in cycles	produced continuously

4.3 **a** Temperature varies slightly during the day, usually rising to a peak in the afternoon and declining at night. **b** Illness often causes a rise in temperature which could be confused with ovulation. **c** **(i)** about 98°F (36.7°C) **(ii)** 98.6–98.8°F (37–37.1°C).
d Temperature stays 'high' due to maintained levels of progesterone. **e** 1st or 2nd of January.

4.4 **a** Enzymes from lysosomes in the secondary oocyte cause the zona pellucida to thicken and separate from the oocyte. The zona forms an impenetrable barrier to sperm.

b If an extra sperm fertilised the egg, three sets of chromosomes (triploidy) would occur. This results in early spontaneous abortion (miscarriage). Fertilisation by several extra sperm would cause polyploidy, also fatal.

c Large wastage is inevitable. For example, sperm may be defective, be killed by acid conditions in the vagina, be unable to penetrate the cervical mucus, or may swim up the wrong oviduct.

4.5 The corpus luteum is the main source of oestrogen and progesterone until the placenta takes over this function three months into pregnancy. Both hormones are essential for maintaining the lining of the uterus and for other functions.

4.6 An excess of testosterone could act by negative feedback in the pituitary gland, blocking the action of FSH and LH and thus switching off its own production (*figure 4.13*). The excess would also have to prevent the normal function of testosterone of switching *on* spermatogenesis in the seminiferous tubules.

4.7 **a**

■ The fact that the male pill makes a man temporarily sterile may threaten the male ego and make men unlikely to use the pill. (Evidence suggests that this will probably not be a major problem.)

■ Fear of being sued if harmful side-effects occur.

■ The consequences of failed contraception affect mainly women, so there is always likely to be more demand for female contraception. Women may not trust men even if the man claims to be 'on the pill'.

■ Side-effects could be a problem. At present there are short-term side-effects such as increased acne, oily skin and weight gain; it can take 4–6 months after stopping the pill for fertility to return; there is concern that long-term side-effects might include heart disease and prostate disease.

b

■ Profit

■ The only other effective contraception for men is the condom and vasectomy.

■ To give greater opportunity to men to share the responsibility for contraception.

■ A spin-off from research could be a better understanding of male infertility.

■ In some cultures women may find it difficult to use contraception for religious or social reasons.

4.8 It prevents implantation rather than fertilisation. RU486 works in this way.

4.9 Difficulty of separation at birth; blood groups of mother and fetus may be incompatible; to prevent the passage of bacteria from mother to fetus; harmful antigen–antibody reponses could occur.

4.10 a Chorion, chorionic villi, blood vessels from the umbilical arteries and vein.
b Endometrium, maternal vein and artery.

4.11 a Oxygen, carbon dioxide, sodium and potassium ions, urea **b** Sodium, potassium and calcium ions, amino acids, iron, vitamins **c** Glucose **d** Water only

4.12 a Energy in the form of carbohydrate or fat; protein; vitamins A, B_1, B_2, B_3, C; folate
b Protein; vitamins A, B_2, C, folate
c Calcium, iron, possibly vitamin D
d Consult *table 4.3*.

4.13 a 20–24. Many people assume the greatest number occurs among teenagers.

b

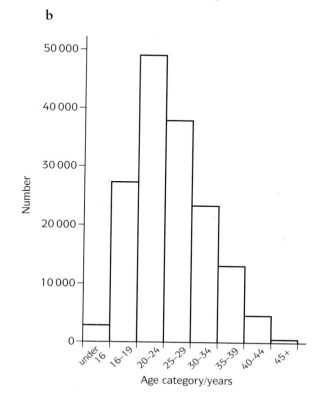

c Total number of women = 160 495

under 16	2%
16–19	17%
20–24	31%
25–29	24%
30–34	15%
35–39	8%
40–44	3%
45 +	0.3%
age not stated	0.007%

4.14 Capacitation must take place first.

Chapter 5

5.1 Day length varies with season, so being sensitive to day length is one way in which a plant can 'know' what time of year it is.

5.2 Temperature.

5.3 Under laboratory conditions, plants could be given short days followed by short nights. SDPs would not flower, but LDPs would. The reverse would be true for long days followed by long nights.

5.4 An absorption spectrum is a graph which shows the relative amounts of light of different wavelengths absorbed by a chemical. An action spectrum is a graph showing the relative effectiveness of different wavelengths of light in bringing about a reaction, such as flowering or photosynthesis.

5.5 Several methods are possible. Leaves could be covered to give them short days while the rest of the plant is exposed to long days. This causes flowering in a SDP. If leaves are exposed to long days while the rest of the plant receives short days, flowering does not occur.

5.6 a Add gibberellin to the starch-agar and use non-embryo halves. Digestion of starch should take place around the non-embryo halves.

b Use starch-agar with non-endosperm halves. No digestion of starch should take place.

5.7 The seeds would only germinate when water was available.

5.8 If seeds with small food reserves are buried in soil, they may have insufficient food reserves to grow to the surface of the soil when they germinate. By being near the surface when germination starts, photosynthesis will start sooner.

5.9 It ensures that the seed does not germinate until it is covered by soil or dead vegetation, when it is more likely to have water and inorganic nutrients.

5.10 Seeds under a vegetation canopy, such as in woodlands and forests, will be prevented from germinating until a break in the canopy allows sunlight through for photosynthesis.

5.11 Phytochrome.

5.12 a Their site of action (target) is always different from their site of synthesis. They must travel in the blood to reach their target. They are a means of communication within the body. b Specific receptor sites exist for the hormones, either in the cell surface membrane or in the cytoplasm. c Made in endocrine glands; small molecules; chemical messengers; travel in blood; act at small concentrations; specific targets. d Good blood supply; secrete hormones directly into blood; ductless.

5.13 a *Luteal progesterone* – initial rise, level, then decline to zero at about 3 months of pregnancy. b *Placental progesterone* – starts to rise as placenta develops (from about 5 weeks) and continues to rise until birth, when it drops to zero. c *Prolactin* – rises sharply after birth. (In fact it starts to rise just before birth, from about 30 weeks of pregnancy.) d *Oxytocin* – rises sharply from just before birth to a peak at birth and then declines to a low level just after birth. Some is produced during lactation. e *HPL* – gradually increases during pregnancy. f *HCG* – rises from time of implantation and then declines from 3 months.

Index (Numbers in italics refer to figures.)